Asset Maintenance
Engineering Methodologies

Asset Maintenance Engineering Methodologies

José Manuel Torres Farinha

CRC Press
Taylor & Francis Group
Boca Raton London New York

CRC Press is an imprint of the
Taylor & Francis Group, an **Informa** business

CRC Press
Taylor & Francis Group
6000 Broken Sound Parkway NW, Suite 300
Boca Raton, FL 33487-2742

First issued in paperback 2020

ISBN-13: 978-0-367-57176-4 (pbk)
ISBN-13: 978-1-138-03589-8 (hbk)

Visit the Taylor & Francis Web site at
http://www.taylorandfrancis.com

and the CRC Press Web site at
http://www.crcpress.com

Contents

Preface

Asset Maintenance Engineering Methodologies, corresponds to the author's global vision of the industrial engineering tools applicable to the management of physical assets corresponding to the state-of-the-art technologies, econometric models, and organizational diagnosis involved in new and future challenges, including the new ISO 5500X standards.

The main chapters correspond to some of the author's research and development, as well as many projects managed by him and implemented in manufacturing industries, transport companies, and hospitals.

The book is a basis to support professionals and researchers in these areas and, obviously, ought to be complemented by the references listed, as well as by academic and professional libraries.

Unlike the author's PhD thesis where he has focused on the equipment life cycle, in other words, on the physical assets life cycle in a whole and integrated way. His approach rather emphasizes on acquisition time to withdrawal, not from the project, as traditionally known as Terology.

It is because of this the book emphasizes the acquisition time, namely from its terms of reference and some standards that can help managers to acquire their physical assets knowing with rigor what is expected of them in terms of availability, maintenance costs, return on investment, and withdrawal time.

Maintenance management is another important area because it is strategic to guarantee the physical asset's life cycle corresponding to that predicted at acquisition time. This approach is used not only at the management level but also with the current technologies, which are fundamental to maximizing equipment availability and, usually, to reducing costs. The author suggests that this last concept ought to be dramatically changed to the *variable investment*, because the physical assets have to be seen not as a cost but as an initial investment at acquisition time and a variable investment during their life cycles.

The present day-to-day technologies, including the Internet of Things and of Internet of Services, sensors, and condition monitoring with predictions based on these technologies, are some other topics that reinforce the main subject.

Organization and management methodologies are very important tools that help one manage assets in a better way. Japanese methods, as well as many others that are used in the most competitive organizations, are also presented in this book.

Finally, some current as well as future technological tools that will dramatically change the way today's physical assets are maintained and managed are presented.

Author

José Manuel Torres Farinha, born in Lisbon, Portugal, graduated in electrical engineering and computers as aggregate. He earned a PhD in mechanical engineering and is a licentiate in electrotechnical engineering (long course with an option—Energy).

Torres Farinha is an auditor of National Defence, Institute of National Defence, employed by NATO and the Common Security and Defence Policy, the Portuguese Atlantic Commission. He is the full professor of mechanical engineering area and a coordinator of MSc in industrial engineering and management at Instituto Superior de Engenharia, Coimbra Polytechnic, Portugal.

He is the president of Scientific Commission of Industrial and Management Engineering, Instituto Superior de Engenharia, Coimbra Polytechnic, Portugal and an integrated member at the Centre for Mechanical Engineering, Materials and Processes (CEMMPRE), Coimbra University, Portugal.

Torres Farinha has been the principal investigator of five PhD theses. He has also supervised 15 MSc theses and 13 are in progress. He has published two books in Portuguese on maintenance management and seven published books to his credit. He also has more than 150 national and international publications.

Torres Farinha regularly participates as plenary speaker at national and international conferences and as reviewer in several international and national scientific and professional journals and also a member of their scientific committees.

Before his academic achievements, he was involved in professional activities such as the following:

- Responsible for general maintenance of hospital facilities in the central region of Portugal, namely lifts, generator sets (emergency), cooking equipment, and cooling and laundry equipment.
- Responsible for electrical networks, including transformer stations.
- Leadership of various maintenance teams for the implementation of planned maintenance in hospitals.
- Design of electrical installations and their specifications, as well as supervision of their execution for the buildings of Regional Health Administration.
- Design of the electrical/electronic projects of lifts for transport of kitchen food in hospitals.
- Design of electronic command systems for glove-washing machines.
- Design of a temperature alarm system (minimum and maximum) for refrigerated medication case.
- Design of a commutation system for diesel–electrical generators.

Torres Farinha was president of one of the biggest higher education institutions in Portugal, having achieved an increase in the size of the institution of more than 50% in terms of number of schools and students, among other equally significant success ratios.

Torres Farinha has to his credit several patents, international and national, namely one related to saving water and another related to rapid cooking.

He has received awards for higher-grade theses and for technological innovations.

1

Introduction

1.1 Background

This book aims to contribute to a new vision of physical asset management and emphasizes several tools to manage the entire life cycle of physical assets. However, because this last concept is not well consolidated yet in professional language, throughout this book, I will use similar concepts, as is the case for the equipment/facilities/machines concepts and the more general concept of maintenance objects.

The objective of this book is to propose real approaches for all phases of a physical asset's life cycle that may be summarized in the followings times and steps:

- t_1—Decision about acquisition
- t_2—Terms of reference
- t_3—Market consultation
- t_4—Acquisition
- t_5—Commissioning
- t_6—Starting production/maintenance
- t_7—Economic/lifespan issues
- t_8—Renewal/withdrawal

The next figure (Figure 1.1) shows the synthesis of all the preceding steps that will be described in this chapter as the global framework of this book.

The cycle starts at time t_1, with the decision about acquisition. This time has much more importance than it has traditionally been given. At this point, aspects like the following are analyzed carefully:

- The physical asset's functions.
- The production levels expected from the asset.
- The estimated budget that may be allocated for its acquisition.
- The geographical location and conditions of asset implantation, which include aspects like the temperature, humidity, and environmental conditions.
- Others may be applicable according to the nature of the activity of the organization.

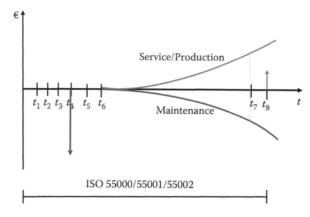

FIGURE 1.1
Stages of a physical asset life cycle.

At time t_1, the decision about acquisition relates directly to the planning requirement of ISO 55001, because it may attend to:

- The asset management objectives and planning to achieve them.
- The strategic asset management plan (SAMP) is the guide to setting asset management objectives.

Time t_2, the terms of reference, is carefully defined. This step must be one of the most relevant for the rest of the asset life cycle. At this point, aspects like the following must be taken into consideration:

- Detailed definition of the asset's functional specifications, namely the following:
  - What are the roles of the asset?
  - How long is the daily operation?
  - What is the risk associated with the use of the asset?
  - What is the risk associated with third parties?
  - Others applicable for each specific situation.
- Detailed service and technical specifications, namely the following:
  - Technical specifications
  - Reliability ratios:
    - Mean time between failures (MTBF)
    - Mean time to repair (MTTR)
    - Mean waiting time (MWT)
    - Others
  - Maintenance specifications
  - Setting of spare parts
- Detailed definition of the documents underlying asset acquisition, like the following:
  - Technical data
  - Operating manual (entry into operation)

- Deployment guide
- List of components and recommended spare parts
- Assembly plan
- Detail plan
- Lubrication plan
- Line diagram
- Logical diagram
- Circuit diagram
- Pipe diagram and instruments
- Drawing of implementation
- Assembly drawing
- Report of the test program
- Certificates

- Detailed definition of the asset's conditions of reception and installation, like the following:
    - In the reception phase, it must be verified whether the delivery complies with the terms of reference specifications and the supplier's proposal.
    - All manuals should be examined and must be complete and match the version of the asset provided.
    - The installation must be coordinated and supervised by a specialist or team of specialists to which the equipment belongs.
    - Before the asset's entry into operation, all licenses and required legal regulatory approvals should be obtained.
- Detailed definition of commissioning conditions:
    - At this stage, it is essential to carry out a set of tests, experiments, and checks to demonstrate and prove that the asset complies with the standards and regulations specified in the terms of reference.
- Other aspects that must be considered are the following:
    - Availability of new technologies
    - Compliance with safety standards or other mandatory regulations
    - Availability of spare parts and the number of years that they are available
    - Obsolescence that can limit the asset's competitive use
    - Guarantees, training, maintenance contracts, and costs associated with the maintenance contract

Time t_3 is for market consultation and has three main steps: (1) invitation of suppliers to make proposals, based on (2) terms of reference, and, finally, (3) reception of proposals from suppliers.

During this phase, suppliers are invited to present proposals that completely respect the terms of reference document, and you should say clearly that only those proposals that correspond to those requisites will be considered. This strategy permits the creation of a consistent framework of analysis to make decisions based on consistent and coherent comparative data.

The next time is t_4, acquisition, in which the followings steps are considered:

- Analysis of proposals from suppliers.
- Analysis of the correspondence between the terms of reference and the proposals.
- Theoretical evaluation, based on econometric models, of the several assets proposed in the suppliers' proposals, taking into considerations aspects like the following:
  - Production capabilities
  - Maintenance costs
  - Return on investment time
  - Others as applicable
- After the previous steps, it is time for the decision about acquisition.

About this point, t_4, ISO 55001 specifies the requirements for the establishment, implementation, maintenance, and improvement of a system for asset management, referred to as an "asset management system."

The time that follows acquisition is t_5, commissioning, which implies the implementation of the next steps:

- Tests and checks to prove the asset complies with standards and regulations as specified in the terms of reference.
- The equipment considered necessary for these tests and measurements must have its calibration certificates updated.
- Tests to show that all equipment, working simultaneously, meets the standards and applicable legal requirements, whether environmental, electrical, security, or other aspects.

Time t_6 is when the production and maintenance procedures start, namely planning, management, and control. Some of the main aspects to be taken into consideration are the following:

- Production planning and management must be considered, like those items referred to in the next bullet points—however, this subject will not be developed at this point or in this book because it is outside of its scope:
  - Enterprise resource planning (ERP) II:
    - Supply chain management (SCM)
    - Customer relationship management (CRM)
    - Business intelligence (BI)
  - Integrated collaboration environment (ICE)
  - Lean management
  - Single-minute exchange of die (SMED)
  - Kanban
  - Supply chain

- Maintenance management is one of the most relevant subjects considered in this book, because the guarantee of the fulfillment of the asset life cycle depends on it. To synthesize this relevance, the following aspects can be considered:
  - Some maintenance management policies:
    - Classification of assets: Class A, B, or C?
    - Total productive maintenance: conventional or other?
    - Scheduled reliability-centered maintenance, risk-based maintenance, no planned maintenance, or other?
    - Condition-based maintenance: predictive?
    - Internet of Things (IoT), Internet of People (IoP), or a mix?
    - Computerized maintenance management system, enterprise asset management, or other?
  - The maintenance organization is a complementary aspect that is intrinsically connected to the previous aspects and in which aspects like the next ones must be considered:
    - 5S/Kaizen/Lean?
    - *Hoshin kanri*/strategic asset management plan?
  - Other relevant aspects that must be taken into account are the following:
    - Which general maintenance standards are necessary?
    - Which specific maintenance standards are necessary?
    - Which maintenance tools must the team know? For example:
      - FMECA (failure mode, effects, and criticality analysis)
      - Fault trees
        - Stochastic
        - Fuzzy
      - Petri networks
      - Markov models
      - Others
  - Another important aspect that must be considered is if the maintenance services are totally or partially conducted externally. In this case, there must be a good terms of reference for outsourcing that must take into account, among others, EN 13269:2007 *Maintenance—Instructions for the preparation of maintenance contracts*, and NP 4492:2010—*Requirements for the provision of maintenance services*. Some aspects that must be considered are the following:
    - Title of the maintenance service
    - Objective of the contract
    - Scope of work required
    - Technical requirements necessary to fulfill service
    - Commercial conditions, like price and conditions of payment
    - Organizational conditions the supplier must accomplish
    - Legal requirements to respect

- Organization of the service
- Service offers that must correspond to those required
- Human resources and respective competences
- Material resources necessary to fulfill the contract
- Management contracts of maintenance services and management control
- Quality program and quality control procedures
- Preparation, planning, and control of the provider
- Engineering studies and requisites of their competences
- Management of materials and parts, including conditions of storage, temperature, humidity, and others
- Safety, health, and environment

The next time that follows is t_7, which corresponds to the evaluation of the economic and/or lifespan life cycle. To evaluate this phase, econometric models may be used that require the values of some variables to evaluate the economic asset life cycle at each time. This analysis must be done for the entire life cycle of the asset because from this analysis, the deviations of the results can be evaluated when compared with the quantitative expectations. Some necessary variables involved are the following:

- Asset acquisition cost
- Cession value
- Operating costs:
  - Maintenance costs—historical costs from working orders
  - Functioning costs—historical costs from energy reports and others
- Inflation rate, according to official sources
- Capitalization rate, according to official sources and monetary conditions of the company

Finally, time t_8 corresponds to the renewal/withdrawal time. If the company follows the asset's life cycle continuously, this is only one more annual exercise. If not, this is an important stage that must be done on time, that is, before the asset reaches the end of its life cycle. As noted in the preceding point, the econometric models may be the same. However, at this point, they emphasize two aspects about the end of the life cycle:

- Economic life
- Lifespan

The variables involved are the same as discussed above:

- Acquisition cost
- Cession value
- Operating costs:
  - Maintenance costs
  - Functioning costs

- Inflation rate
- Capitalization rate

As synthesized above, the life cycle of an asset runs from time t_1 until t_8 with the contents described in this chapter. Transverse to all these times is the ISO 55000/1/2:2014 Standard—*Asset management*, which has aspects that accompany all asset life cycles.

In fact, according to ISO 55000:2014, "Asset management enables an organization to examine the need for, and performance of, assets and asset systems at different levels. Additionally, it enables the application of analytical approaches towards managing an asset over the different stages of its life cycle (which can start with the conception of the need for the asset, through to its disposal, and includes the managing of any potential post disposal liabilities)."

In this book, the ISO 55001 standard will be referred to. The main requirements that an organization must fulfill are the following:

- Context of the organization
- Leadership
- Planning
- Support
- Operation
- Performance evaluation
- Improvement

These requirements will be referred to in several chapters of the book in order to mesh the engineering methodologies with the physical asset requisites of the norm and several tools that will be introduced throughout the book.

1.2 Book Structure

Multiple aspects described in the previous points are developed in several chapters of this book, as explained next:

- *Chapter 2—Terology Activity*: This chapter explains the terology concept versus the terotechnology concept to frame the importance of the physical asset life cycle from its project to its withdrawal. Nowadays, the new physical asset management requirements and the increased competitiveness of organizations give a higher relevance to the concepts the book author supports in his research and development and also his professional activity since many years ago.

- *Chapter 3—Physical Asset Acquisition and Withdrawal*: The two most symbolic and relevant points of a physical asset are its acquisition and withdrawal times, because they involve the evaluation of several variables, like the acquisition, maintenance, and functioning cost, as well as the inflation and market money cost, among others. Other important aspects to be managed are related to the elaboration of the terms of reference, which is the most important document that determines all asset life cycles, from acquisition until withdrawal.

- *Chapter 4—Diagnosis of Maintenance State*: One of the most important aspects in the asset life cycle is its maintenance that must be done according to the best methodologies that maximize its availability. However, to make it possible to achieve this, it is necessary to have an adequate maintenance organization and management. One way to evaluate whether this happens is to make a periodic diagnosis of the state of the maintenance organization or, if it is the beginning of the process, to make an initial diagnosis to make it possible to restructure the maintenance organization and implement the procedures, starting work with a good level of organization and management.

- *Chapter 5—Maintenance Management*: Maintenance management usually represents the largest time activity in an asset life cycle. There are several approaches that can be used according to the specificity of each asset. However, there are transversal aspects of management, as is the case for maintenance planning, control, work order management, outsourcing management, reports, and key performance indicators (KPIs), among others, that are strategic for good maintenance management. These are some of the main aspects that are presented in this chapter.

- *Chapter 6—Maintenance Resources*: Maintenance activity is possible only if it uses adequate resources, namely human resources, materials/spare parts, and tools. Obviously, financial resources are subjacent to all resources. The internal dimensioning of the necessary resources depends on the balance between internal maintenance and outsourcing.

- *Chapter 7—Integrated Systems for Maintenance Management*: Nowadays, all management support must be paperless. The concepts based on "e-" are historical, because all administrative procedures are based on digital documents, the technicians receive work orders and equipment manuals on their tablets and smartphones, and so on. It is based on these principles that this chapter presents the main modules of computerized maintenance management systems (CMMSs) and, in a more general way, enterprise asset management (EAM) systems that manage all physical asset life cycles.

- *Chapter 8—Expert Systems for Fault Diagnosis*: One of the weakest points of maintenance activity of equipment is when a fault occurs. Service manuals and the history retrieved from working orders are good sources of information. However, this process can be much improved through the use of expert systems (ESs) for fault diagnosis. There are historic case studies that show its importance and current solutions. But, in many cases, the specificity and cost of these systems implies that most organizations do not use them. Regardless of this, and with the advent of the newest software tools and their integration within the CMMS, ESs have an important role to play in the day-to-day support of maintenance management.

- *Chapter 9—Maintenance 4.0*: Nowadays, the concept of Industry 4.0 is in vogue. However, this concept does not add any intrinsic value, having merit because it emphasizes the integration of all technologies that, in the present day, can contribute together to help adequately manage assets. The Internet of Things, sensing, condition monitoring, predicting through adequate software tools, and many others technologies like these are presented and discussed in this chapter from the perspective of Maintenance 4.0.

- *Chapter 10—Forecasting*: Maintenance planning must correspond to the most relevant part of maintenance interventions, being scheduled or conditioned. This chapter presents some time series algorithms that are very important to help in managing maintenance planning, both periodic and aperiodic, including condition monitoring with or without prediction.

- *Chapter 11—Maintenance Logistics*: One aspect that is often ignored is maintenance logistics. This problem is evident in many situations, like the following: paths between the workshop and the warehouse, travel between the organization and commercial warehouses, and travel among several locations of the assets when they are dispersed geographically. When these problems happen, the associated costs must be carefully considered, and logistic optimization can achieve important cost reduction and asset availability improvements.

- *Chapter 12—Condition Monitoring*: Condition monitoring is more and more important for many assets, since this approach increases availability with reduced costs and has been improved by monitoring through the IoT. There are many variables that can reduce equipment "health," and it is necessary to identify each one that affects its condition. However, there are many variables and solutions that are well consolidated in the practices of organizations, like vibrations, oil analysis, temperatures, electrical voltage and currents, and effluents, among many others. These are the main subjects that are explored in this chapter, which opens the discussion for a more general approach to condition monitoring according to each real asset.

- *Chapter 13—Dynamic Modeling*: In many situations, it is necessary to understand details of asset functioning in the case of fault diagnosis, reliability analysis, or equipment improvement, among others. To reach these objectives, there are some important tools that can be used, like Petri networks, Markov and hidden Markov chains, and fault trees, both stochastic or fuzzy. These are the main aspects that are described in this chapter in order to enlarge the reader's knowledge of the tools that can help in asset life cycle management.

- *Chapter 14—3D Systems*: When a technician performs a maintenance intervention, many times he or she needs to access manuals and schema. Even when the information is in digital format, it is usually static, especially the recent ones, as well as technical drawings. If these are 3D and dynamic, and it is possible to manipulate them, then interventions, namely the most complex ones, can be improved in quality, minimizing errors and reducing intervention time and, obviously, costs. These are the main aspects discussed in this chapter, as well as the connection to the future through augmented reality markerless systems.

- *Chapter 15—Reliability*: Asset management implies good maintenance management, which implies a good reliability approach, and this is more or less emphasized. This chapter describes the most relevant statistical distributions applied to reliability with exercises to more clearly show its potential in day-to-day use to aid maintenance activities from the perspective of reliability-centered maintenance (RCM) or any other model.

- *Chapter 16—Management Methodologies*: Management methodologies are fundamental to make possible a good/excellent level of maintenance management and, obviously, asset management. Several methodologies are presented in this chapter, like 5S, poka-yoke, the PDCA (plan, do, check, act) cycle, A3 maps, the GUT (gravity, urgency and trend) matrix, and Lean maintenance, among others. These tools are fundamental to organizing and managing asset maintenance in particular and the asset life cycle in general. These are the main subjects presented in this chapter as consolidated methodologies in current competitive organizations.

- *Chapter 17—Maintenance Standards*: Nowadays, the asset life cycle has many national and international standards that can help a lot with its management. This chapter emphasizes two types of standards, ISO 5500X and maintenance standards. Obviously, links to other international standards that can help in asset management are presented, according to the specificity of each one.

- *Chapter 18—Maintenance Project Management*: Physical asset management implies decisions like the following: renewal, periodic stoppage for general maintenance, and major maintenance interventions. In these situations, one must have good projects and planning in order to accomplish the objectives within the time foreseen. Some of the usual good tools that can be used are the Gantt map, the program evaluation and review technique (PERT), and the critical path method (CPM), which are useful, easy to manage, and can be used with a lot of software, both proprietary and open source, that each organization can use easily. Additionally, these software tools can be connected directly or through a spreadsheet to the CMMS or EAM used by the organization. These are the main subjects discussed in this chapter that correspond to some of the most traditional project management tools but that continue to be useful to many organizations.

- *Chapter 19—Maintenance Training*: Asset management in general and asset maintenance in particular are very difficult activities, which is reflected in training activities, even for well-trained professionals. These activities imply a continuous upgrade in training for all professionals involved, which implies high costs for it. Additionally, training is usually outside the organization, which implies additional costs. New technologies can help to minimize these problems and, in many cases, increase the quality of training, as is the case with new technologies like artificial vision, augmented reality, and, in the near future, holography. These are the main subjects presented and discussed in this chapter, according to the state of the art.

- *Chapter 20—Terology Behind Tomorrow*: This chapter presents the author's vision of the future of asset management, based on the current state of the art and provisional evolution. If today can be represented by Physical Asset 4.0, this chapter will present Asset Management 4.1.

2

Terology Activity

2.1 Background

Terotechnology is defined as "the technology of installation, commissioning, maintenance, replacement, and removal of plant machinery and equipment, with feedback on the operation and design thereof and on related subjects and practices. Terotechnology is the maintenance of assets in an optimal manner. It is the combination of management, financial, engineering, and other practices applied to physical assets such as plants, machinery, equipment, buildings, and structures in pursuit of economic life cycle costs. It is related to the reliability and maintainability of physical assets and also takes into account the processes of installation, commissioning, operation, maintenance, modification, and replacement. Decisions are influenced by feedback on design, performance, and cost information throughout the life cycle of a project. It can equally be applied to products, as the product of one organization often becomes the asset of another" (Husband, 1976).

The terotechnology concept appeared in the early 1970s in the United Kingdom, and it constitutes a deep content about physical assets. It was a very important concept in this country, being the target of the following British Standards:

- BS 3811:1993, *Glossary of terms used in terotechnology*
- BS 3843-1:1992, *Guide to terotechnology (the economic management of assets)—Part 1: Introduction to terotechnology*
- BS 3843-2:1992, *Guide to terotechnology (the economic management of assets)—Part 2: Introduction to the techniques and applications*
- BS 3843-3:1992, *Guide to terotechnology (the economic management of assets)—Part 3: Guide to the available techniques*

In the same way and at the same time, the Japanese concept of total productive maintenance (TPM), which is based on the following five points (Takahashi, 1981), appeared:

1. To establish objectives that maximize the effectiveness of the assets
2. To establish a comprehensive productive maintenance system that fully covers the asset's life cycle
3. To get the involvement of all departments, such as planning, operations, and maintenance
4. To get the participation of all members, from senior management to workers
5. To strengthen the motivation of the staff, creating small autonomous groups of productive maintenance

According to Takahashi (1981), the terotechnology and TPM concepts are similar, although the latter has additional concerns with the motivation of staff, as is typical in the Japanese industrial culture.

This terotechnological maintenance concept combines two aspects:

1. The technology of maintenance that applies engineering knowledge appropriate to the nature of the interventions on the equipment
2. The maintenance management that "takes care" of assets, as well as the actions and resources required to maintain them

From this formulation, but removing the technological component reflected in the word *tero(techno)logy*, the new concept of terology was introduced (Farinha, 1994), defined as "the combined utilization of operational research techniques, information management, and engineering, with the objective of accompanying the life cycle of facilities and equipment; it includes the definition of specifications of purchase, installation, and reception, and also the management and control of its maintenance, modification, and replacement and its accompanying in service, too."

From the point of view of implementation, terology must be analyzed using its strategic, tactical, and operational aspects:

- Strategic—Is related to new procurement policies, definitions of methodologies of management, control of the life cycle of physical assets, the necessary human and material resources, and their replacement.
- Tactical—Is related to the standardization problems of the physical asset, its maintainability and reliability, cost control of the several resources involved, and personnel training, among others.
- Operational—Is related to maintenance itself, that is, to guarantee the normal operation of physical assets by planning and controlling interventions. These, in turn, provide technical and economic data, which provide several indicators necessary to evaluate the performance of the defined strategy.

This is the extensive vision of physical asset management that supports the global approach of this book, which is based on maintenance management itself, but with an enlarged vision of its life cycle.

2.2 Concept of Terology vs. Maintenance

The terology concept has its core in maintenance activity, as seen in the previous section.

According to EN 13306 *Maintenance terminology*, maintenance is the "combination of all technical, administrative, and managerial actions during the life cycle of an item intended to retain it in, or restore it to, a state in which it can perform the required function."

This definition of maintenance emphasizes its importance during the life cycle of an asset, but omits the extreme points of an asset's life: its acquisition and all activities that must be fulfilled in this phase, and its withdrawal, which represents the other extreme of the life cycle and its final phase.

All assets' phases, including maintenance management; organizational, administrative, and econometric models; and engineering methodologies, among others, are relevant to global physical asset management.

It is because of this that the author introduced the terology concept, which includes the whole physical asset life cycle, including the definition of specifications of purchase; installation and reception; management and control of its maintenance, modification, and replacement; and its accompanying in service, too.

In fact, the acquisition phase has implications for the whole life cycle of an asset, namely its availability and maintenance cost, including resources involved; its depreciation; and, finally, the replacement decision or, if it is the case, renewal and a new life cycle.

2.3 Terology as a Multidisciplinary Issue

As stated before, the concept of terology implies an enlarged view of maintenance and physical asset management. This concept can be decomposed into several elements such as the following:

- Operational research
  - Operational research makes use of mathematical, statistical, and algorithmic models to aid the decision-making. These strands are present throughout the maintenance management process and, in particular, in the algorithms that are embedded in the information systems that support maintenance management.
- Information management
  - Information management is a fundamental pillar in monitoring and maintenance control, being transversal to its development in several aspects. It can be limited to management of the maintenance with data deferred in time or can include aspects such as online reading of the data of the assets, including automatic prediction of maintenance interventions. It is also decisive in aspects such as fault diagnosis and the evaluation of key performance indicators.
- Engineering
  - Engineering is present in most maintenance activities in several specialties, namely electrotechnics, electronics, mechanics, electromechanics, and mechatronics, among others.
- Reliability
  - Reliability is another crosscutting aspect of maintenance activity, such as in intervention planning, resources, and, in particular, spare parts and in all phases of the physical asset life cycle.
- Invoicing
  - Invoicing is present in maintenance activity, particularly when it is outsourced, when part or all of the interventions are allocated to direct and indirect customers of the equipment, and when purchasing spare parts, among others.

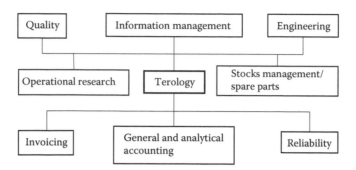

FIGURE 2.1
Interdisciplinary maintenance (terology).

- General and analytical accounting
  - General and analytical accounting is the area of the organization that acts as a barometer of the financial control, being in this case crucial for the monitoring of the execution of the maintenance budget. It is also relevant for the definition of the dates of amortization and reintegration of the assets, as well as a support for asset renewal.
  - When entering the data of the equipment into the information management system, it should be immediately indexed to the accounts, either through inventory codes or analytical accounting.
- Stock management/spare parts
  - Spare parts management is crucial to the proper execution of maintenance interventions. When a maintenance action is taken, it is essential that all spare parts be available and that the logistics associated with them be optimized so that the time associated with their availability and application is minimal.
- Quality
  - Finally, maintenance is intrinsically linked to the organization's quality system as an integral part of it and due to the relevance of the evidence that organizations must demonstrate. These can be in the ambit of the industrial environment in the quality of its maintenance to continue guaranteeing its accreditation and, as a consequence, the confidence of its stakeholders.

Figure 2.1 illustrates several maintenance areas in the broad perspective of terology, with the objective of highlighting some aspects of the interdisciplinary of this activity, which, because of its diversity, represents some of the reasons that justify its complexity and difficulty of management.

2.4 Terology and the Environment

Better maintenance, aided by information systems, helps increase assets' life cycles, reducing effluent emissions and energy consumption. These are only some reasons to say that maintenance or, in other words, terology, helps the environment.

Nowadays, economic recession, environmental degradation, and the imminent catastrophe of the planet are enough reasons to look at maintenance cut as the way to minimize the impact of human beings on nature through fewer effluent emissions; redesign physical assets with the objective of not taking more resources from nature; and help to maintain, or to increase, the quality of life for everybody. In the ambit of a circular economy, this can be the differentiating variable. This is the real challenge that must be overcome.

The economy has to be redesigned, always considering that the new economy will be an ecology economy, one that will allow human beings to live in harmony with nature, preserving it, returning to nature what it has given us and what we have transformed but, after it has been used extensively, returning it, with no aggression, in an ecological way—maybe the circular economy will design a new future.

Terology has the potential to help manage assets designed by humans in a better way, minimizing or at least annulling the effects of those assets on nature. It is because of this that we believe terology can have a decisive contribution to a sustainable future.

The economy and well-being of the nineteenth and twentieth centuries were supported by intensive consumption of natural resources. These were of all types, namely forests, minerals, and fossil resources.

The result of this approach of managing nature with the goal of creating new goods for well-being was and is the degradation of the planet by several ways, such as the following:

- Humans took resources from the interior and surface of the earth.
- Humans transformed natural resources into artificial goods using energy, initially from forests and coal and, nowadays, from petroleum and gas; the use of renewable sources remains insignificant.
- The functioning of many goods consumes a lot of energy.
- Heat cannot be sent outside the earth, and it causes the planet to become, over time, hotter and hotter, like a sauna.
- Additionally, and just importantly, artificial goods send a lot of chemical products into the atmosphere, and to the surface and the interior of the earth, that are contaminating the planet.
- Also important are wars, forest fires, and other accidents that degrade the planet.

All this degradation is possible because the evolution of knowledge was exponential, and this has been the bible of humanity. Science has always believed that the humanity could manage, control, and always find solutions to all of nature's problems.

The results that we are feeling in our lives show that it was not humans who dominated nature, but that humans are ill, like the planet, because of inadequate use of knowledge. However, and finally, it will be nature that will win the war, as always.

The main objective of all kinds of production is to be effective or, at a minimum, efficient. This means that the goal is to take more and more from the earth, from the sea, or from any place where there are natural resources that can add value to the market.

The main objective seems to be to empty all resources so efficiently as possible, but when the danger of really emptying a certain resource seems imminent, then it is easy to make a natural park, a bank of seeds, or something similar, believing that on an uncertain day in the future, that resource will be rehabilitated. However, as if this weren't bad enough, nobody cares if some animal and plant species disappear.

Where do maintenance, terology, and physical asset management enter?

Maintenance and terology have been seen as a way to help keep production as high as possible. But, in this way, it is seen as a cost that, as a consequence, diminishes the profits of companies. This implies that maintenance has been managed as an activity less important and less valued than other activities in companies, namely production.

Nevertheless, the competences and knowledge involved in the maintenance area by technicians are very high and some of the most eclectic within a company. Then, why not take this potential to guarantee a better and more sustainable future? And how?

Under the maintenance theme of physical assets, it assumes several approaches, namely the guarantee of reliability, renewal, replacement, and all associated activities, including purchase and withdrawal—in other words, the maintenance approach of terology.

However, the econometric models that are included within algorithms have the principle that natural resources are infinite, or almost, and the life cycle of physical assets is seen without any importance by the environment.

Then, a problem arises: it is currently necessary to change the economic paradigm to an ecological economy paradigm. This may signify that it would be necessary to deconstruct the present economy and reconstruct it, but what price is it necessary to pay for that?

The price is, first of all, political. It is necessary to assume the cost of survival! However, it is necessary to make the change quickly, but carefully, in order not to have many victims. The majority of business is directly or indirectly dependent on petroleum; thus, it is necessary to make changes taking the following into account:

- Do not construct new buildings unless it is extremely necessary; in this case, the materials ought to be ecological and the new buildings should necessarily be autonomous in terms of energy, water, and so on.
- Adapt present buildings to be more efficient from the point of view of energy, at least in order to reach the objective of being autonomous.
- The same is true for water, namely recycling used water and storing rainwater.
- Using solar panels to heat water.
- Using and reusing ecological bags to transport purchases.
- Using ecological public transport that has been transformed from combustion engines to ecological ones.
- Using private transport only when necessary and transforming it from combustion engines to ecological ones.
- Buying a new car only when necessary; and, obviously, it must be ecological.
- New materials will only be ecological.
- Planting trees in cities in as many areas as possible! All cities will have a balance between green areas and constructed areas.
- All citizens are responsible for making these changes in a well-defined period of time.

As can be seen, maintenance or, in other words, terology, is present at all points of the change that can help humanity to believe in its future and in the future of the planet (Farinha, 2009).

2.5 Related Concepts

Terology as a multidisciplinary concept accompanies the asset's life cycle from acquisition until withdrawal. However, maintenance management is one of its most important components, which implies an emphasis on the main aspects under its role.

The concept of maintenance was described above and will be discussed in the next point. However, it is relevant to emphasize multiple aspects of organizations related to maintenance activity, as described in detail above. But, first of all, it is important to define the ambit of maintenance management and, obviously, of the assets, or, in other words, the facilities and equipment involved and the relative importance of each one.

It is usual to classify assets into several categories, for example, categories A, B, and C following a Pareto classification. Usually, the A category corresponds to the most important one on which the main production depends, one whose risk level is very high, or one that is otherwise important for the organization. The B category is relevant, but much less so than the A category. The C category is one that is not relevant for production or, obviously, the organization's income.

The definition of the asset's categories permits one to get to the next step, which is to define the most adequate type or types of maintenance management for each category. This means that each organization may have more than one type of maintenance policy, according to the asset specificity.

The global solution is the one that allows achievement of the best result, which implies the maximization of the availability at the most rational cost and at the minimum risk.

2.5.1 Maintenance

As described before, according to EN 13306 *Maintenance terminology*, maintenance is the "combination of all technical, administrative, and managerial actions during the life cycle of an item intended to retain it in, or restore it to, a state in which it can perform the required function."

Regarding the maintenance concept, it is also pertinent to talk about the maintenance types that can be synthesized in the following way:

- Planned maintenance
  - Systematic/scheduled
  - Conditioned
- Unplanned maintenance

Regarding planned maintenance, interventions follow a previously established program that aims at the following goals:

- To prevent faults or malfunctions and balance the maintenance workload
- To suit the interventions to the asset production program
- To prepare resources in advance to make interventions more economic and cost effective

Under systematic planned maintenance, the interventions follow a program that is intended to be executed periodically, with the intervals measured in a given unit of time or another variable of use that translates into the asset's functioning.

Under the condition of monitoring maintenance, actions are performed according to the "health" status of the asset, which may also be a malfunction if this is the previously planned condition. In general, several variables may be associated with the asset, measured in a given unit that, when it reaches certain limits, gives rise to an intervention.

Unplanned maintenance includes all unplanned interventions.

The previous concepts are also compatible with the commonly used terms of preventive maintenance and corrective maintenance. Planned, systematic, or conditional actions are obviously preventive, but may also include corrective maintenance work defined at the time of the intervention.

An unplanned maintenance intervention will be, in most cases, corrective and supposedly curative, but sometimes palliative. At the time of implementation of these interventions, preventive (planned) maintenance actions may also be included, if justified at the time of intervention, in order to optimize resources and increase the availability of the asset.

Under this context, the terminology included in EN 13306 defines maintenance types and strategies in the following way:

- Preventive maintenance
  - Maintenance carried out at predetermined intervals or according to prescribed criteria in order to reduce the likelihood of damage or degradation of the operation of an asset.
- Scheduled maintenance
  - Preventive maintenance carried out according to a pre-established schedule or according to a defined number of units of use.
- Systematic maintenance
  - Preventive maintenance carried out at pre-established time intervals or according to a defined number of units of use but without previous control of the asset "health."
- Condition monitoring maintenance
  - Preventive maintenance based on the monitoring of the functioning of the asset and/or significant parameters of such operation, integrating the resulting actions.
    - Note—Monitoring of the operation and some asset parameters can be carried out according to a schedule, by request, or continuously.
- Predictive maintenance
  - Conditional maintenance carried out in accordance with the extrapolated forecasts of the analysis and evaluation of significant parameters of the asset degradation.
- Corrective maintenance
  - Maintenance carried out after the detection of a fault and intended to restore an asset to a state in which it can perform a required function.
- Remote maintenance
  - Maintenance of an asset carried out without a physical access to it by the staff.

- Deferred maintenance
  - Corrective maintenance that is not performed immediately after the detection of a fault state, but is delayed according to certain maintenance rules.
- Urgent maintenance
  - Corrective maintenance that is performed immediately upon the detection of a fault state to avoid unacceptable consequences.
- Maintenance in operation
  - Maintenance carried out during the time at which the property is in operation.
- On-site maintenance
  - Maintenance performed in the place where the asset works.
- Operator maintenance
  - Maintenance performed by a user or operator of the asset.

2.5.2 Total Productive Maintenance

The concept of total productive maintenance is attributed to Nippondenso, a company that created parts for Toyota. However, Seiichi Nakajima is regarded as the father of TPM because of his numerous contributions to TPM.

The TPM concept can be understood as an important umbrella that covers a lot of concepts to manage maintenance and rationalize it by taking into account the specificity of physical assets.

TPM can be considered a system for maintaining and improving the integrity of production and quality systems through equipment, processes, and employees that add business value for an organization. TPM focuses on all equipment being in top working condition to avoid breakdowns and delays in manufacturing processes.

The objectives of TPM are to increase the productivity of a plant with a rational investment in maintenance.

In order to make TPM effective, the full support of the total workforce is required. This should result in accomplishing the goal of TPM: to enhance the volume of production, employee morale, and job satisfaction.

The main objective of TPM is to increase the overall equipment effectiveness (OEE) of plant equipment. TPM addresses the causes of accelerated deterioration while creating the correct conditions between operators and equipment to create ownership.

The five pillars of TPM are mostly focused on proactive and preventive techniques for improving equipment reliability, as said before and referred to again here:

1. Establishing objectives that maximize the physical asset's efficiency
2. Establishing a global system of productive maintenance that fully covers the asset's life cycle
3. Obtaining the involvement of all departments, such as planning, operations, and maintenance
4. Obtaining the participation of all members, from the top management to the workers
5. Strengthening staff motivation by creating small autonomous groups of productive maintenance (Figure 2.2).

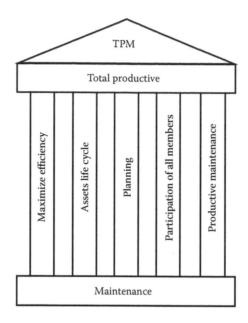

FIGURE 2.2
TPM pillars.

Based on these pillars, productivity can be increased through the increase of availability. The implementation of TPM involves the following steps:

- Initial evaluation of TPM level
- Introductory training and information spread about TPM
- Training of TPM committee
- Development of master plan for TPM implementation
- Stage-by-stage training for employees and stakeholders on all five pillars of TPM
- Implementation of the TPM preparation process
- Establishing the TPM policies and goals, and development of a road map for TPM implementation

Additionally, many times, the concept of total quality management (TQM) in conjunction with total productive maintenance is considered the key operational activity in the quality management system.

TQM and TPM are often used interchangeably. However, though TQM and TPM have a lot of similarities, two different approaches by several authors are considered. TQM attempts to increase the quality of goods, services, and customer satisfaction by raising awareness of quality concerns across the organization. TPM is based on five keystones:

1. The product
2. The process that allows the product to be produced
3. The organization that provides the settings needed for the process to work
4. The leadership that guides the organization
5. The commitment to excellence throughout the organization

TQM focuses on the quality of the product or service, while TPM focuses on the physical assets used to produce the products by preventing equipment breakdown and improving the quality of functioning of the equipment.

Considering this approach to adequate availability according to the type of physical assets, in the ambit of TPM, planned maintenance ought to be implemented obeying the specificity of physical assets and their situation. This means that planning can be: scheduled; RCM; condition-based maintenance (CBM); risk-based maintenance (RBM); or other according to the asset specificity, but taking into account a TPM approach.

2.5.3 Reliability-Centered Maintenance

The concept of reliability-centered maintenance corresponds to an industrial management approach focused on the identification and establishment of maintenance improvement policies and capital investments, leading for managing risks of equipment failures more effectively. This concept is defined by American Society of Automobile Engineers (SAE) technical standard JA1011, called *Evaluation Criteria for RCM Processes*.

RCM corresponds to an engineering framework that allows a broad conception of this activity, which views maintenance as a means of maintaining the functions that the user requires for physical assets in its operational context. In fact, physical assets may have several failure rates, with diverse behaviors, the bathtub curve being the most iconic. Figure 2.3 shows some failure curves, but some others can be added. Chapter 15 of this book deals with this subject in more detail.

As a discipline, RCM allows external monitoring of equipment in order to evaluate, predict, and, in a general way, understand the operation of the assets from the perspective of their reliability. This is assumed in the phases of the RCM process:

1. In the first phase, the operational context of the asset is identified, which is usual when a failure mode, effects, and (criticality) analysis (FME[C]A) is done.

2. In the second phase, RCM is applied according to the most appropriate maintenance procedures for the failure modes identified by the FMECA analysis.

When it is complete, the list resulting from maintenance procedures is grouped so that the periodicity of the activities is rationalized and handled from planned work orders (WOs). Finally, RCM is maintained throughout the operational life cycle of the assets, with

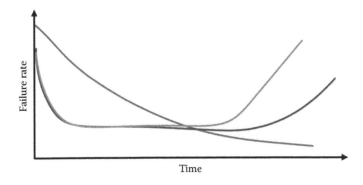

FIGURE 2.3
Some failure rate curves.

planned maintenance being permanently supervised and subject to a systematic review, adjusted based on equipment history.

According to the Department of Defense (DoD, 2011), "RCM is used to determine what failure management strategies should be applied to ensure a system achieves the desired levels of safety, reliability, environmental soundness, and operational readiness in the most cost-effective manner. In the context of RCM, this can mean identifying various maintenance actions. For example, one of the most useful products of an RCM analysis is the identification of technically defensible proactive maintenance tasks such as condition monitoring, scheduled restoration, and scheduled discard tasks. RCM can yield other results that also contribute significantly to the safe and reliable operation of assets. These can include design modifications, changes to a training program, identification of new operating and emergency procedures, or modifications to technical manuals."

According to the same document, RCM is based on the following principles:

- Seeks to preserve a desired level of system or equipment functionality.
- Manages the life cycle from design through disposal.
- Seeks to manage the consequences of failures, but not prevent all failures.
- Identifies the most applicable and effective maintenance tasks or other actions.
- Is driven by (listed in order of importance) safety or a similarly critical consideration such as environmental laws, the ability to complete the mission, and economics.
- Acknowledges the design limitations in the operating context. Maintenance can sustain the inherent level of reliability within the operating context over the life of the asset.
- Is a continuous process that requires sustainment throughout the life cycle. RCM uses design, operations, maintenance, engineering, logistics, and cost data to improve operating capability, design, and maintenance.

The emphasis on this last concept gives relevance to the asset life cycle: ". . . life-cycle management tool that should be applied from design through disposal." The tendency of all concepts within the ambit of asset management is to manage facilities and equipment in a broad view, not only with a technical management approach in an operational context.

2.5.4 Risk-Based Maintenance

The concept of risk-based maintenance is based on the five core elements of the risk management process, which are the following:

1. Identification
2. Measure
3. Risk level
4. Estimation
5. Control and monitoring

The methodology underlying the RBM concept is based on the integration of the reliability aspect with the level of risk, having the objective to obtain optimal maintenance planning, implemented from the following steps:

1. First—The probable scenarios of equipment failure are formulated, being the most likely subject of a detailed study.
2. Second—A detailed analysis of the consequences for the selected situations is made.
3. Third—The study of these cases is made through fault trees or a similar tool to determine the probability of failure.
4. Fourth—The level of risk is evaluated by combining the analysis of the consequences with the probabilities of failure.
5. Fifth—The calculated risk is compared to find the acceptable risk criteria.

The periodicity of maintenance interventions is obtained by minimizing the estimated risk.

2.5.5 Other Concepts

The Pareto principle (also known as the 80/20 rule, the law of the vital few, or the principle of factor sparsity) states that, for many events, roughly 80% of the effects come from 20% of the causes. Essentially, Pareto showed that approximately 80% of the land in Italy was owned by 20% of the population. Pareto developed the principle by observing that about 20% of the peapods in his garden contained 80% of the peas.

It is a common rule of thumb in business that 80% of sales come from 20% of clients.

Mathematically, the 80/20 rule is roughly followed by a power law distribution for a particular set of parameters, and many natural phenomena have been shown empirically to exhibit such a distribution.

Why is Pareto distribution so important in this context?

Almost no organizations need to have planned maintenance for all their assets. Usually, it is a rule of thumb to classify the assets into A, B, and C classes, as noted in a section above. The A-class assets are the ones that correspond to the production, risks, or similar assets the organization depends on. The B class contains the assets with intermediate importance. The C class contains the irrelevant ones.

This classification is very important for maintenance asset management. In fact, organizations don't plan the management of all their assets in the same way.

The assets that are directly related to production activities and or have a severe risk level must be managed as class-A assets. The intermediate ones are managed as B-class assets and the irrelevant as C class.

This stratification is extremely relevant, because the importance of assets in any organization depends on its specificity and, obviously, it implies strategies and management approaches according to its nature that are related to its importance for the organization's production.

Then, after defining the A, B, and C asset classes, the next step is to decide about life cycle cost (LCC) management, including the maintenance policy, that can be one of those referred to above and/or scheduled planning or condition-based maintenance.

Scheduled maintenance is the most common way to plan maintenance and, probably, the first planning tool in history. It can be described as a periodic intervention on the physical asset based on a well-defined variable. This can be real time (calendar), hours of operation, or another variable that systematically triggers a planned working order to intervene with the asset.

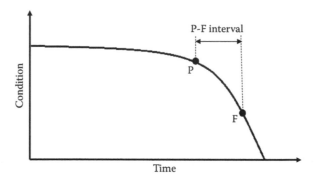

FIGURE 2.4
P-F curve and P-F interval.

Condition-based maintenance is maintenance planning that involves monitoring the equipment condition (condition monitoring; CM), usually predicting equipment failure.

This maintenance strategy aims to extend equipment life, increase availability, lower operating costs, and decrease the number of maintenance interventions. Unlike periodic maintenance, where services are based upon scheduled intervals, CBM relies upon actual machine health to dictate when and what intervention is required.

The CBM approach tends to be the most common in industry unless other planning is required because of specific reasons. It requires knowledge of CM techniques, and a high investment is needed for data measurement tools and databases. However, with the advent of the Internet of Things and the price of sensors lowering, this type of planning continues to increase more and more.

CBM is usually complemented by a prediction tool in order to maximize the interval between interventions. This approach is usually described by the PF curve. The curve shows that as a failure starts manifesting, the equipment deteriorates to the point at which it can Possibly be detected (P). If the failure is not detected and mitigated, it continues until a "hard" Failure occurs (F). The time range between P and F, commonly called the P-F interval, is the window of opportunity during which an inspection can possibly detect the imminent failure and address it. P-F intervals can be measured in any unit associated with the exposure to the stress (running time, cycles, miles, etc.) (Figure 2.4).

3

Physical Asset Acquisition and Withdrawal

3.1 Background

From the time a decision is made about physical asset acquisition, all the aspects inherent to its life cycle, including the operating conditions, until its withdrawal must be duly considered.

The acquisition of a physical asset involves a previous assessment of its expected life, the functioning of which is usually measured in real time, working hours, or another variable, according to its specificity and type of operation.

This chapter presents the aspects related to acquisition, both in administrative and LCC management and withdrawal, and also the relation between LCC and ISO 55001.

3.2 Purchase of Physical Assets

The care taken in the process of acquisition of a physical asset will be reflected throughout its entire life cycle. Thus, it is extremely important to take into account aspects such as (Farinha, 1994):

1. Terms of reference
2. Physical asset selection
3. Reception, installation, and commissioning
4. User training

The next step is the operation of the physical asset, including the necessary interventions, planned and unplanned, and, finally, its withdrawal, which is the last point of its life cycle.

During the phase of purchase and operation, EN 13460:2009 *Documentation for maintenance* should be taken into account, since it defines the general guidelines for:

- "The technical documentation to be provided with a physical asset, at the latest before it is put into service in order to support their maintenance..."
- "Information/documentation to be established during the operational phase of the physical asset in order to support the maintenance needs..."

From this perspective, this standard describes the key documents and gives additional information necessary to the process of acquisition and operation of physical assets, such as the following:

- Documents from the preparation phase:
  - Technical data
  - Operation manual (operating instructions)
  - Implementation guide
  - List of components and recommended spare parts
  - Assembly plan
  - Detailed plan
  - Lubricating plan
  - Line diagram
  - Logical diagram
  - Circuit diagram
  - Diagram of pipes and instruments
  - Layout drawing
  - Implantation drawing
  - Report of test program
  - Certificates
- Documents from the operational phase:
  - Document index
  - Physical asset registration (basic data of equipment)
  - Historical record of maintenance interventions
  - Work orders
  - List of reference of spare parts
  - Diagram of cause and effect
  - Historical record of values of parameters
  - Control table of MTBF and MTTR
  - Planning sheet
  - Programming sheet
  - Production planning
  - Sheet of availability and physical asset utilization
  - Human resources activity
  - Historical record of other resources
  - Historical record of maintenance costs
  - Organizational chart of the company
  - Reviews by direction of the quality management system, objectives, and maintenance policies
  - Procedures for maintenance contracts and their annexes

- Maintenance contracts and their annexes
- Procedures to review the causes of critical failures
- Procedures to assess the time of maintenance operations for critical failures (MTTR, MTBF)
- Procedure to control documents and maintenance data
- User privileges of the information maintenance system
- Manual of the information maintenance system
- Approved service providers
- Procedure to evaluate service providers
- Procedure to issue purchase orders for maintenance materials (spare parts)
- Purchase orders for maintenance materials
- Procedure to verify materials purchased
- Verification of purchased materials
- Procedure to control the materials supplied by the customer
- State of the products supplied by customers
- Procedure to identify materials
- Procedure for traceability
- Procedures to control maintenance activities
- Procedure to control a general maintenance activity
- Procedures concerning critical maintenance activities
- Procedure to monitor and test materials (during downtime and operation)
- Incorporation of materials not checked for urgent maintenance
- Procedure to calibrate critical test equipment
- Procedure to identify the test equipment that affects the effectiveness of the means of production (critical test equipment)
- Calibration records of the critical test equipment
- Procedure to identify, document, and so on the conditions of noncompliance of products due to maintenance
- Procedure to prevent and correct actions
- Procedures to handle, storage, pack, preserve, and deliver
- Control of maintenance records
- Procedures to plan and execute maintenance internal audits
- Maintenance internal audits
- Procedure to follow up on corrective actions of an internal audit
- Procedure to identify the requirements for training
- Registration of staff skills and training files
- Procedure to monitor, verify, and report maintenance interventions by outsourcing
- Maintenance services provided
- Procedure to monitor the application of statistical techniques
- Procedures enabling compliance with laws and regulations

The accomplishment of the preceding aspects is a guarantee that the purchase phase of a physical asset will provide a strong background for the next phases of the physical asset according to the expected LCC.

3.2.1 Terms of Reference

The delivery of a product or service to the customer with quality assurance, according to the terms of reference (TR), must be increasingly a cultural pattern in the organization's activity and less a differentiating factor.

Additionally, the quality assurance performed by internal resources or by outsourcing is strategic to the organization's success, enabling equipment performance at the required level. Moreover, in some cases, there is a lack of knowledge concerning the state of the art on the part of some maintenance customers and a lack of quality of some maintenance service providers, which contributes to the low quality of many interventions and an increase of potential risks (Nunes, 2012).

The TR is a strategic element for all organizations, particularly in public institutions, both for acquisition of physical assets and for contracting maintenance services. In the first case, it focuses on the issues inherent to the expected life cycle of assets. These can be measured through the costs of the asset life cycle and economic KPIs like return on investment (ROI), among others. Regarding the acquisition of maintenance services, these must comply with national and international standards, balancing rights and obligations between customer and provider, with the objective of maximizing the physical asset's availability.

The TR can be considered "a contractual document describing what is expected of the Supplier by the Client, being the first the entity chosen by the Client to do the work, in terms of time, quality and costs fixed by the latter in accordance to the contract, and the Contractor is the Client who is buying the service/equipment" (Kioskea.net).

In the case of equipment purchase, LCC analysis is a technique that has been widely used as an engineering and management tool: "The life cycle cost of an asset, by definition, is the sum of all capital spent in support of that asset, since its design and manufacture, through the operation to the end of its useful life" (White and Ostwald, 1976).

In the case of purchase of services, there is a set of rules that must be followed to ensure their quality, both from the customer's and supplier's point of view, such as NP 4492:2010—*Requirements for the provision of maintenance services*, which is focused on the customer and aligned with ISO 9001, also including environmental and safety requirements.

3.2.2 Terms of Reference for Physical Asset Acquisition

The purchase of a physical asset, due to the need of a new one or from the need to replace one because it has reached the end of its life, involves analyzing which physical assets fit the life cycle most appropriately to the investment. For this purpose, it is necessary to evaluate aspects such as (Farinha, 2011):

- Availability of new technologies
- Compliance with safety standards or other mandatory requirements
- Availability of spare parts
- Obsolescence that can limit the asset's competitive use
- Guarantees, training, maintenance contracts, and costs associated with the maintenance contract

The acquisition of a new piece of equipment or renewal of an old one implies costs and income for many years. It is a widespread and incorrect practice to forget these aspects and decide on the purchase just based on the lowest price (cost of initial investment). Given that resources are always scarce, usually the decision is based on the lowest purchase price and not the LCC. However, the best decision ought to be based on the latter, which means "the lowest cost over the life cycle" and not "the lowest investment" (Assis, 2010).

An asset management system should enable companies to maximize the value of their products and services through the optimization of their life cycles. These objectives must be based on decisions involving the appropriate selection of equipment, adequate operation and maintenance and, at the end of the asset's life, its renewal or withdrawal.

The LCC is an analysis technique that has been used as an engineering tool (e.g., supporting the project and acquisition) and as a management support tool (e.g., cost analysis).

As noted before, equipment LCC is the sum of all capital spent in support of an asset from its design and manufacture through operation until the end of its useful life. It is understood that the lifetime goes until the withdrawal of the equipment.

The LCC of an equipment that can be significantly higher than the value of the initial investment and, in many cases, it can be set in the design phase.

The main goal of LCC in equipment management is to support decisions based on analysis of alternatives by the estimation of its total costs during its life cycle. This calculation allows emphasis on the equipment total costs, aiding selection of the best solution.

It is based on the preceding aspects that are relevant to analyzing the acquisition terms of reference document where the information should be mentioned, such as that specified in the following items:

I. *Functional specifications*: The functional specifications must be included in a document that describes all features that the physical asset must have—the functions that customers and users want from the equipment define its desired functionality. For its definition, aspects such as the following must be defined:

- What is the role of the asset?
- How long is the daily operation?
- What is the risk associated with the use of the asset?
- What is the risk associated with third parties?
- Others

When purchasing an asset, the TR should specify the maintenance aspects, spare parts, and human resources by specialty, with the objective that the asset will ensure the reliability and maintainability expected.

II. *Service and technical specifications*: The acquisition of any physical asset should involve an exhaustive definition of technical specifications, in particular its suitability for their operational context.

According to EN 13306:2007 *Maintenance terminology*, reliability is the "ability of an asset to fulfill a required function under given conditions for a given time." Therefore, it is important to define the technical specifications that suppliers must fulfill for assets during the expected life with well-defined reliability and availability. Based on this, the aspects that must be considered are the following:

- Technical specifications
- Reliability ratios

- MTBF
- MTTR
- MWT—This corresponds to the average waiting time between the identification of the fault and the beginning of corrective maintenance intervention.

- Maintenance specifications
- Setting of spare parts

III. *Documents underlying equipment acquisition*: EN 13460:2009 *Maintenance— Documentation for maintenance* presents the general guidelines for the technical documentation to be provided with an asset before it is put into service. The goal is to support its maintenance and for the documentation to be established during the asset operational phase in order to support the maintenance needs. When an asset is ordered from the supplier, those documents and information must be explicitly defined on the order and correspond to the terms of reference:

- Technical data—Corresponds to the manufacturer's specifications.
- Operating manual (entry into operation)—Corresponds to the technical instructions for obtaining the proper functioning according to the specifications and safety conditions.
- Deployment guide—Contains maintenance and technical instructions to maintain and restart asset functioning to fulfill its required function.
- List of components and recommended spare parts—Full list of modules to complement the asset and the spare parts for its maintenance.
- Assembly plan—Refers to the drawings showing the arrangement of the components for an asset.
- Detail plan—Refers to the drawings with the parts list to allow disassembly, repair, and assembly of assets.
- Lubrication plan—Refers to the drawings with the position of each lubrication point, with data and lubrication specifications.
- Line diagram—A general diagram or schema that shows the circuits of power, pneumatic, hydraulic, or other systems. This diagram should be included in the circuits of the distribution panel.
- Logical diagram—A system control diagram to clarify the general structure of the system.
- Circuit diagram—The main power diagram and control circuits.
- Pipe diagram and instruments—Fluid pipelines and control circuits.
- Drawing of implementation—Drawing showing the location of all assets in the area concerned.
- Assembly drawing—Drawing with all areas of a particular installation.
- Report of the test program—Reception or commissioning report showing that the asset complies with the specifications.
- Certificates—Certificates relating to specific safety and legal provisions for assets (lifting equipment, steam boilers, pressure vessels, etc.).

Another element that should be delivered with the equipment is the quality documentation, which describes all materials used in the equipment manufacturing and all tests (cracking, radiography, liquid penetrant, etc.) carried out before considering the equipment fit for delivery.

3.2.3 Reception and Installation of Equipment

After acquiring an asset, it is necessary to proceed with its reception and installation. During the reception phase, it should be carefully observed if the delivery complies with the TR specifications and the supplier's proposal. After this aspect is verified, the manuals should be examined; they should be complete and should match the version of the equipment provided (Farinha, 2011). A simple inspection by the technicians does not represent the final reception. Tests should be done to evaluate the quality of the assets. The contractual responsibility of the supplier continues until the final acceptance of the equipment.

During the installation of the physical asset, some sectors should intervene, such as the maintenance department, users, and inventory department. It is also necessary to verify if the installation conditions comply with the specifications. Information and knowledge about the installation site are extremely important: both the physical and environmental conditions that affect this space. If you do not take these aspects into account, you may incur potential damage, interference with other devices or users, or even the loss of the equipment warranty, which is an aspect that should be taken into account during the elaboration of the TR. The installation must be coordinated and supervised by a specialist or a team of specialists.

In both phases, the asset should be marked with a code and inserted into the information system, as should its maintenance plans. Adequate resources to ensure its life cycle must be provided.

Before the asset enters into operation, all licenses and required legal regulatory approvals should be obtained.

3.2.4 Commissioning

At the stage of commissioning, it is essential to carry out a set of tests, experiments, and checks to demonstrate and prove that the asset complies with standards and regulations as specified in the TR. The equipment considered necessary for these tests and measurements must have its calibration certificates updated.

All the necessary tests must be made to show that all equipment, working simultaneously, meets the standards and applicable legal requirements, whether in environmental, electrical, security, or other aspects.

All these tests shall be done according to the national and international standards and regulations applicable.

3.2.5 Terms of Reference for Acquisition of Maintenance Services

Nowadays, activities related to maintenance have a significant relevance in operating costs. The cost of maintenance services is constantly increasing in the budgets of companies looking for these to optimize their operations and minimize risks. The terms of reference document for acquisition of maintenance services is a strategic document for quality and global asset management.

The preparation of this document implies some data collection for each department involved, taking into account the nature of the equipment. At the end, a TR document is created about the maintenance requisites for each equipment that is target of the TR.

The TR must contain all the information about the maintenance requisites for each equipment. It is because of this that the documentation requested at the time of the asset purchase, namely the service manual, is so important. A vital point to mention is related to the fact that the contractor must provide the spare parts or not, or if it is necessary the installation of external equipment to carry out interventions and an indication of who will be responsible for those costs, among others.

In a maintenance service contract, there are important elements that must be met to guarantee its success and that it protects both parties. EN 13269:2007 *Maintenance— Instructions for the preparation of maintenance contracts* provides a list of elements that a maintenance service contract should contain. Some elements of that agreement are the following:

  I. *Title*—Identifies the parties and the contract.

 II. *Objective*—Defines the general intentions of the parties and contract objectives. These are the key points, especially for long-term contracts.

III. *Scope of work*

   • Operating site—Describes the site where the asset to be maintained is located.

   • Content—Refers to the tasks (what and when) to be performed by the maintenance service provider and, always necessary, the ones excluded from the scope of the contract. Some tasks may include: procedures that must be performed; expected outcomes; measurable targets for the results; tools, equipment, and techniques that have to be applied; security requirements; physical assets involved; qualification of workers; operating conditions of use; and so on.

   • Time—Sets the time period during which the tasks have to be performed, such as the minimum and/or maximum time between the request and the start of work, or the date or period within which tasks must be performed.

IV. *Technical requirements*

   • Verification—This point includes specific information about the conditions/ requirements that must be met for acceptance of the work by the contractor: who checks it (or it may be by delegation), how the verification is done (procedures), when the check is made, what is verified, and so on.

   • Technical information—It should include a list of all relevant technical information that must be exchanged between the service provider and the contractor (EN 13460:2009—*Maintenance documents for maintenance*), set by whom (service provider, contractor), how (registration means, procedures, error correction), when (date, time), and what information (work performed, installed spare parts, overtime, delays, repaired damage, etc.) should be exchanged.

   • Spare parts/materials and consumables—It should contain information such as responsibility for providing these, property, quality required, supplier (from origin or not), location and responsibility for storage (stored by the service provider or contractor; ordering procedure of spare parts; consequences for the contractor for not ordering or ordering late; changing the type of spare parts and versions of integrated software, if applicable; and consequences for the contractor for not ordering new software versions), availability, and delivery time.

V. *Commercial conditions*

- Price—Specification for financial compensation to the service provider for carrying out the maintenance tasks.
- Payment terms—Specification of the conditions of payments required under the contract.
- Guarantees—Definition of contractor's rights and obligations of the supplier of services in the event of noncompliance, as provided in the contract.
- Penalties/assumed damage—Specification of penalties and damages assumed to be paid in case of noncompliance with the contract.
- Insurance—Description of insurance that may be required by the contract or by law, which party will be responsible for providing it, and its procedures.

VI. *Organizational conditions*

- Conditions for implementation—This should include the list of services and resources to be provided by each part under the contract: services; storage space; energy, water, and so on; special tools; spare parts; necessary measures to allow the maintenance personnel to work, travel, have a place to stay, and obtain work permits; working and environmental conditions; integration of maintenance tasks to be performed by the contractor; and requirements for working schedule.
- Health and safety—It should describe hygiene and safety regulations required by law, specific safety regulations of the parties involved in the contract (e.g., site emergency plan and risk analysis), necessary training and allowances for service provider workers, and supply of personal protective equipment (clothing, vaccinations, health certificates).
- Environmental protection—It should contain provisions for handling, segregation, and waste removal and pollution prevention (liquids, gases, solid materials, and noise).
- Security—It should include provisions for special requirements for equipment, safeguarding information and data systems, documentation of security measures, authorization/access licenses, and confidentiality—protecting information.
- Quality assurance—It should include measures to ensure quality: the quality management system of the service provider, requirements for audits, experience and qualifications of personnel, transfer of knowledge between the parties, quality indicators, and quality plan (as defined in the ISO 9000 series).
- Supervision/management—It should define the methodology related to the management control.
- Records—It must describe relevant events that may affect contractual agreements, which must be recorded: *what*—work done, installed spares, time spent, overtime, anomalies, obstacles, delays, and so on; *by whom*—service provider, contractor, subcontractors; *when*—date, exact time and duration of the occurrence; and *how*—recording means, procedures, and evidence.

VII. *Legal requirements*

- At this point, it should introduce and define the rights of ownership and existing rights of use, confidentiality as protection and information security, legal responsibility definitions for damage caused by one of the parties or their workers during the contract period, definitions of the reasons and procedures for withdrawal, and so on.

The maintenance performed by internal resources of the organization or through outsourcing by specialized external suppliers has a key role in the success of the organization: ensuring that the equipment performs at the expected level. In addition, a lack of knowledge about the state of the art by some purchasers of maintenance services and the lack of quality of some maintenance service providers contribute to the failure of interventions and increase the risk of maintenance activity.

To help to solve these problems, in Portugal, there is the standard NP 4492:2010—*Requirements for the provision of maintenance services*, which is strategic to the success of maintenance services. The requirements of maintenance service providers should meet the point five of this standard related to customer satisfaction and also the service provider. Some requirements are the following:

I. *Organization*—The requirements for maintenance service providers offer a variety of contracts involving multiple specialties and work in any place and at any time. Maintenance services should be performed by companies with an appropriate structure for material and human resources, organization, and information systems.

II. *Services offer*—The provider of maintenance services should, clearly, completely, and in detail, define the object and scope of its services offered. For each individual type of service, it should define the necessary expertise, demonstrating how to guarantee it, documenting its support, and ensuring its access. It should ensure the quality of the proposed service, demonstrating its experience, portfolio, and results of its activity. So, it ought to keep an up-to-date list of references of services and documentation certifying the qualification or certification to provide such services.

III. *Human resources*—The provider of maintenance services must set up a human resources framework to ensure the quality of service with competence, based on education, specific training, appropriate qualifications, and professional experience.

IV. *Material resources*—The maintenance service provider facilities should meet the needs of management and operation of its portfolio of services. For equipment, tools, and instruments, the maintenance services provider must meet the needs of operability of its services offer. This includes the accomplishment of technical and quality standards, the maximizing of the productivity of services, the size of the staff team, and the reduction of the risk to safety and overall costs, applying its own resources and or those of subcontractors. Regarding customer property, the service provider should take care of it while it is under its control or being used by the company. It shall identify, verify, protect, and safeguard customer property provided for use or incorporated in the service.

V. *Management maintenance service contracts and management control*—This point refers to the establishment between the contractor and maintenance services provider of a framework in the ambit of the contracted services. This item refers to EN 13269:2007 *Maintenance—Instructions for the preparation of maintenance contracts*. In management control, the responsibility of the maintenance service provider is related to the establishment and monitoring of the performance indicators of the services provided (EN 15341:2009 *Maintenance—Maintenance performance indicators* [KPI]) to diagnose eventual deviations from the objectives, to implement preventive/corrective actions, and to conform to management and budgetary control.

VI. *Quality program*—The provider of maintenance services should apply appropriate methods for monitoring and, when applicable, measurement and analysis of the services performed. The company must define methods to obtain data to monitor customer satisfaction through pragmatic KPI.

VII. *Preparation, planning, and control of the provider*—The provider shall ensure the availability of all tools and equipment necessary for the contractualized services. Also, the provider must ensure that all laws and regulations related to the exercise of the tasks necessary for the execution of work are fulfilled, including those related to hygiene, health, and safety. The company service provider shall monitor compliance with the implementation of assistance, according to the planning and work program, correcting deviations when they happen. The report on interventions and their control will be provided to the contractor in accordance with the agreement between it and the service provider.

VIII. *Engineering studies*—The company may use technical engineering work to safeguard the efficiency of its work.

IX. *Management of materials and parts*—The service provider must establish objectives and targets with respect to: shipping; stock management; storage; cost; responsibilities of stakeholders in the material and part management process; decision-making criteria for its own stock or of supplier contracts, with a delivery guarantee in a well-defined time; a procedure for adequate stock management; and a suitable location for the storage of materials and parts.

X. *Safety, health, and environment*—The provider of maintenance services should check local security conditions for carrying out the work and ensure the availability of necessary means of protection. Personnel involved in the tasks of maintenance services must have training and adequate knowledge of the specific risks of the tasks to be performed.

3.3 Maintenance of Physical Assets

The maintenance of physical assets implies the definition of their importance, individually or by family, following an approach like an ABC or Pareto analysis or another method according to their specificity.

It also implies defining what type of methodology to follow—for example, scheduled maintenance, condition monitoring, with or without prediction, risk-based maintenance, and/or corrective maintenance.

These decisions must be according to the situation of each organization, namely if its assets are new, old, or in the middle age of their LCC. Each company must be analyzed individually.

There are several mathematical tools that may be applied, but the data and situation of each company are different from each other.

3.4 Resources and Budgeting

Maintenance activity must use resources to be accomplished, namely human, materials, tools, and financial. Usually, this last one is left to the financial department of the company.

The first resources referred to are discussed in several chapters of this book. In this section, financial resources, including the maintenance budget, are briefly discussed.

To design the maintenance budget, some variables must be considered, like the following:

- Human resources, internal
- Spare parts and other materials
- Tools
- Outsourcing
- Structural costs

How does the company design the budget?

It is fundamental to have a CMMS up to date to be possible to design the budget. From the maintenance plans, the company can know the following data:

- Total planned working hours for all assets and, as a consequence, total human resources cost
- Total spare parts necessary and, as a consequence, total cost
- Total tools necessary and, as a consequence, their costs, including depreciation

From a historic point of view, the following data can be evaluated:

- Total nonplanned working hours for all assets and, as a consequence, total human resources cost
- Total spare parts necessary for nonplanned WO and, as a consequence, their total cost
- Total tools necessary for nonplanned WO and, as a consequence, their costs, including depreciation

From the difference between the total necessities and internal resources, it is possible to define the total outsourcing necessary.

From the cost of internal resources plus the cost of outsourcing, it is possible to evaluate the main part of the maintenance budget.

Additionally, it is necessary to add depreciation costs, namely buildings and other physical assets associated with the maintenance department.

Finally, the costs of energy, water, and others must be added. All these added costs constitute the global maintenance budget.

The absolute costs are specific for each company, according to its specificity and dimensions. However, in relative terms, they may be compared with similar companies in order to evaluate their performance. KPIs like return on investment and return on assets (ROA) are some examples of ratios that can be used to evaluate the company performance, including its maintenance department. However, bringing in information data from sales and production performance, among others, cannot be forgotten. The maintenance department is strategic for the company's success, but it needs the other departments to work well.

3.5 Econometric Models

The econometric models used in this book take into account several variables, such as:

- Acquisition cost
- Cession value
- Operating costs
  - Maintenance costs
  - Functioning costs
- Inflation rate
- Capitalization rate

The values of most of the above variables are obtained from the asset's history, with the exception of the cession value. In this case, it is necessary to have the market value for each piece of equipment, which may be difficult for many assets. Alternatively, several types of depreciation, such as the following, can be simulated (Farinha, 2011):

I. *Linear depreciation method*—The decay of the equipment value is constant over the years.
II. *Sum of digits method*—The annual depreciation is not linear but less than exponential.
III. *Exponential method*—The annual depreciation is exponential over the equipment life.

I. *Linear depreciation method*: This method considers that the decay of the equipment value is constant over the years and is calculated as follows:

$$d = \frac{C_0 - R}{N} \tag{3.1}$$

where:
d—Annual quota of depreciation
C_0—Original value of the equipment
R—Residual value of the equipment at the end of N periods of time
N—Lifetime corresponding to R

The value of the equipment V_n in a period n shorter than N is given by:

$$V_n = C_0 - n * d \tag{3.2}$$

II. *Sum of digits method*: In this case, the annual depreciation is not linear—it is calculated as follows:

$$SD = \frac{N(N+1)}{2} \tag{3.3}$$

where:

N—Lifetime corresponding to R

R—Residual value of the equipment at the end of N periods of time

$$d_n = \frac{N-(n-1)}{SD}(C_o - R) \tag{3.4}$$

d_n—Annual quota of depreciation
C_0—Original value of the equipment
n—Devaluation period

The value of the equipment V_n in a period n shorter than N is given by:

$$V_n = V_{n-1} - d_n \tag{3.5}$$

where V_{n-1} is the value of the equipment in the previous period. For the period 1, that is, V_0, the value coincides with C_0.

III. *Exponential Method*: The exponential method uses an annual exponential decrease in depreciation over the life of the equipment. The calculation formula is as follows:

$$V_n = C_0(1-T)^n \tag{3.6}$$

where:

C_0—Original value of the equipment
T—Annual depreciation rate
V_n—Value of equipment in year n
n—Devaluation period

If the residual value is known, the exponential rate of depreciation using the following formula can be determined:

$$T = 1 - \sqrt[N]{\frac{R}{C_0}} \tag{3.7}$$

where:

N—Lifetime corresponding to R

If the residual value R is null, the above formula cannot be applied, so an exponential depreciation rate must be defined.

Table 3.1 and Figure 3.1 illustrate an example of calculation for the three previous methods in the case of depreciation of equipment with an initial cost of 1200 monetary units (MU) and a residual value of 70 MU after five years.

TABLE 3.1

Simulation for Equipment Depreciation

Period	Acquisition Cost	Linear Depreciation	Sum of Digits	Exponential
0	1200	1200	1200	1200
1		974	823	680
2		748	522	385
3		522	296	218
4		296	145	124
5		70	70	70

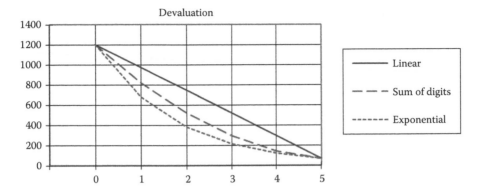

FIGURE 3.1
Different depreciation methods.

The calculations for the first two periods for each method are shown below:

- Linear depreciation method

$$C_0 = 1200 \text{ MU}$$

$$R = 70 \text{ MU}$$

$$N = 5$$

$$d = \frac{1200 - 70}{5} = 226$$

$$V_1 = 1200 - 1 * 226 = 974 \text{ MU}$$

$$V_2 = 1200 - 2 * 226 = 748 \text{ MU}$$

- Sum of digits method

$$C_0 = 1200 \text{ MU}$$

$$R = 70 \text{ MU}$$

$$N = 5$$

$$SD = \frac{5(5+1)}{2} = 15$$

$$d_1 = \frac{5 - (1-1)}{15}(1200 - 70) \cong 376.7$$

$$V_1 = V_0 - d_1 = 1200 - 376.7 \cong 823 \text{ MU}$$

$$d_2 = \frac{5 - (2-1)}{15}(1200 - 70) \cong 301.3$$

$$V_2 = V_1 - d_2 = 823 - 301.3 \cong 522 \text{ MU}$$

- Exponential method

$$C_0 = 1200 \text{ MU}$$

$$R = 70 \text{ MU}$$

$$N = 5$$

$$T = 1 - \sqrt[N]{\frac{R}{C_0}} = 1 - \left(\frac{R}{C_0}\right)^{1/N} = 1 - \left(\frac{70}{1200}\right)^{1/5} = 1 - 0.0583^{0.2} \cong 0.434$$

$$V_1 = 1200(1 - 0.434)^1 \cong 680 \text{ MU}$$

$$V_2 = 1200(1 - 0.434)^2 \cong 385 \text{ MU}$$

3.6 Life Cycle Costing

The LCC of an asset is the sum of all capital spent in support of that asset since its design and manufacture through operation until the end of its useful life. The scope of this book has only a marginal relation with the design and manufacture of physical assets, but has a strong approach from their acquisition until withdrawal. It is in this ambit that life cycle costing is managed and makes use of the preceding and next math models for asset management.

The sum of all recurring and nonrecurring costs over the full life cycle of a physical asset includes the purchase price, installation cost, operating costs, maintenance and upgrade costs, and remaining value at the end of ownership or its useful life.

One interesting question that can be posed is the evaluation of an LCC in situations like the following:

1. LCC evaluation for a new physical asset
2. LCC evaluation for a used physical asset
3. LCC evaluation at any time

The first situation implies simulating the costs along the predicted life and the remaining value at withdrawal time according to market conditions.

The second situation usually implies using the historical asset and knowing the devaluation over time with the objective of evaluating the most economic time to cession or the lifespan.

The third situation aims to evaluate and simulate the LCC at any time, which implies the use of historical information about the time from acquisition until the present, and also the simulation of future costs and cession values. These situations imply making backward and forward monetary corrections according to the apparent rate.

The algorithms and methods presented in this chapter and its sections answer all these questions.

3.7 Withdrawal

The withdrawal time for a physical asset can have several reasons, like the following:

1. When it reaches the end of its economic life
2. When it reaches its lifespan
3. When it reaches its obsolescence time
4. When it reaches its environmental impact limit

This book deals with the two first situations, which are analyzed in detail in the next sections.

When one of the preceding situations is reached, it is time to evaluate which is the best solution:

1. To sell the physical asset according to the market value and acquire a new one
2. To renew the asset

The reason to choose between these two options ought to be based on the evaluation of the asset's next LCC. For the first option, the analysis may be based on the methodologies proposed in this book. For the second option, the renewal costs must be evaluated and, if they are lower than the cost of a new asset and the new LCC is similar to that of a new asset, then this option is probably better than the first, and vice versa.

3.8 Methods for Replacing Assets

To evaluate the replacement time, in addition to the information mentioned in previous sections, some more data must be introduced, as will be described in this section.

The asset can be replaced by several criteria, as explained in the previous section. For the financial criterion, one usual approach is the economic life, which allows one to determine the most rational time to withdraw that minimizes the average total cost of operation, maintenance, and capital immobilization. Another method commonly used is the lifespan, which states that the life cycle ends when the maintenance costs surpass the maintenance costs plus capital amortization of a new equivalent asset.

However, despite being possible to know the market depreciation for some equipment, two other variables should also be taken into account:

- The capitalization rate, called i
- The inflation rate, called ϕ

These rates are related as follows:

$$i_A = i + \phi + i * \phi \tag{3.8}$$

where i_A is the apparent rate.

For example, for an inflation rate of 4% and a capitalization rate of 11%, the value of the apparent rate will be:

$$i_A = 0.04 + 0.11 + 0.04 * 0.11 = 0.1544$$

3.8.1 Determination of the Economic Life for Replacement

There are several methods for determining the economic cycle of replacement equipment. The most common are:

  i. Uniform annual income method (UAIM)
 ii. Minimizing the total average cost method (MTACM)
iii. MTACM with reduction to the present value method (MTACM-RPV)

  i. The UAIM makes use of the following data:

- Asset acquisition value
- Cession annual values (evaluated according to the above methods or the market values)
- Maintenance and operation costs over the years
- Apparent rate

The uniform annual income U of the possession of equipment is given by:

$$U = \frac{i_A(1+i_A)^n}{(1+i_A)^n - 1} * \sum_{j=0}^{n} \frac{X_j}{(1+i_A)^j} \qquad (3.9)$$

where:
i_A—Apparent rate
n—Period for which U is calculated
X_j—Maintenance and operating costs for the period j

where:

$$P = \sum_{j=0}^{n} \frac{X_j}{(1+i_A)^j} \qquad (3.10)$$

This is called the present value of X_j values and

$$U' = \frac{i_A(1+i_A)^n}{(1+i_A)^n - 1} \qquad (3.11)$$

This is called the factor of capital recovery of a uniform series of payments.

The lower U value calculated indicates the respective period (multiple of the year) in which the asset must be replaced. This value is equivalent to a minimum amount the equipment costs annually.

The following example (Table 3.2 and Figure 3.2) illustrates a situation for determining the most rational asset replacement time—its acquisition values, exploration, and cession are indicated in italics. An apparent rate of 8% was considered.

TABLE 3.2

Table to Determine the Amount of Annual Income

Years	0	1	2	3	4	5
Acquisition (X_0)	*1200.00*					
Exploration (X_j)		*740.00*	*770.00*	*840.00*	*1000.00*	*1200.00*
Present value (*P*)		1885.19	2545.34	3212.16	3947.19	4763.88
Cession (y_j)		*880.00*	*640.00*	*440.00*	*250.00*	*70.00*
Present value (*P'*)		814.81	548.70	349.29	183.76	47.64
P − P'		1070.37	1996.64	2862.87	3763.43	4716.24
Annual income (*U*)		1156.00	1119.65	*1110.89*	1136.26	1181.21

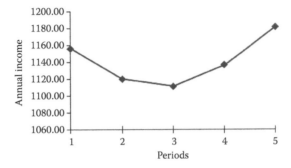

FIGURE 3.2
Evolution of the value of annual income.

When the asset reaches withdrawal time, it will be possible to recover some capital through its sale. This implies that it is necessary to subtract the amount of the sale from the costs after its correction to the present value ($P − P'$). It is based on this result that the value of the uniform annual income is determined. The resulting replacement interval has three periods. The calculations for the first two periods are illustrated next.

$$P_1 = \left[\frac{1200}{(1+0.08)^0} + \frac{740}{(1+0.08)^1}\right] = \left[1200 + \frac{740}{1.08}\right] = 1885.19$$

$$P_1' = \frac{880}{(1+0.08)^1} = 814.81$$

$$U_1 = \frac{0.08*(1+0.08)^1}{(1+0.08)^1 - 1}(P_1 - P_1') = 1.08*(1885.19 - 814.81) = 1156.00$$

$$P_2 = \left[\frac{1200}{(1+0.08)^0} + \frac{740}{(1+0.08)^1} + \frac{770}{(1+0.08)^2}\right] = \left[1200 + \frac{740}{1.08} + \frac{770}{1.17}\right] = 2545.34$$

$$P_2' = \frac{640}{(1+0.08)^2} = 548.70$$

$$U_2 = \frac{0.08*(1+0.08)^2}{(1+0.08)^2 - 1}(P_2 - P_2') = 0.56*(2545.34 - 548.70) = 1119.65$$

ii. The MTACM determines the lowest average cost of equipment ownership and the period in which it occurs that corresponds to the most rational replacement time. The costs of capital and the inflation rate are not considered. The calculation procedure is as follows:

$$C_n' = \frac{\sum_{i=1}^{n} C_{Mi}}{n} \tag{3.12}$$

$$C_n'' = \frac{V_A - V_{Cn}}{n} \tag{3.13}$$

$$C_n = C_n' + C_n'' \tag{3.14}$$

where:

C_{Mi}—Maintenance and operating costs in the year i

V_A—Acquisition value in year 0

V_{Cn}—Cession value in year n (calculated according to the methods set out above or the actual market value)

C_n—Total average cost

The following example (Table 3.3 and Figure 3.3) uses the data from the previous example, and also indicates, in italics, the acquisition, operation, and disposal values.

TABLE 3.3

Table for Determining the Average Total Cost

Years	0	1	2	3	4	5
Purchase (V_A)	*1200.00*					
Exploration (C_{Mi})		*740.00*	*770.00*	*840.00*	*1000.00*	*1200.00*
C_n'		740.00	755.00	783.33	837.50	910.00
Cession (V_{Cn})		*880.00*	*640.00*	*440.00*	*250.00*	*70.00*
C_n''		320.00	280.00	253.33	237.50	226.00
$C_n' + C_n''$		1060.00	**1035.00**	1036.67	1075.00	1136.00

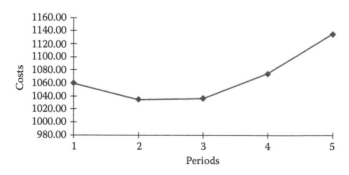

FIGURE 3.3

Evolution of the value of the average total cost.

It can be seen in the table or figure that the replacement interval has two periods. The calculations for the first two periods are illustrated below:

$$C'_1 = \frac{740}{1} = 740$$

$$C''_1 = \frac{1200-880}{1} = 320$$

$$C_1 = 740 + 320 = 1060$$

$$C'_2 = \frac{740+770}{2} = 755$$

$$C''_2 = \frac{1200-640}{2} = 280$$

$$C_2 = 755 + 280 = 1035$$

iii. The MTACM-RPV calculation procedure is similar to the previous ones, except here the cost of capital and the inflation rate are considered. Several maintenance and cession values over the years are reduced to the present value, according to the following procedure:

$$C'_n = \frac{1}{n}\sum_{i=1}^{n}\frac{C_{Mi}}{(1+i_A)^i} \tag{3.15}$$

$$C''_n = \frac{V_A - (V_{Cn}/(1+i_A)^n)}{n} \tag{3.16}$$

$$C_n = C'_n + C''_n \tag{3.17}$$

where:

C_{Mi}—Maintenance and operating costs in the year i

V_A—Acquisition value in year 0

V_{Cn}—Cession value in year n (calculated according to the methods set out above or the actual market value)

C_n—Reduced total average cost to the present value

The following example (Table 3.4 and Figure 3.4) uses the data from the previous example, with the acquisition, operation, and disposal values also indicated in italics.

TABLE 3.4

Table to Determine the Total Average Cost Reduced to Present Value

Years	0	1	2	3	4	5
Acquisition (V_A)	*1200.00*					
Exploration (C_{Mi})		*740.00*	*770.00*	*840.00*	*1000.00*	*1200.00*
Present value (C'_{Mi})		685.19	660.15	666.82	735.03	816.70
C'_n		685.19	672.67	670.72	686.80	712.78
Cession (V_{Cn})		*880.00*	*640.00*	*440.00*	*250.00*	*70.00*
Present value (V'_{Cn})		814.81	548.70	349.29	183.76	47.64
C''_n		385.19	325.65	283.57	254.06	230.47
$C'_n + C''_n$		1070.37	998.32	954.29	*940.86*	943.25

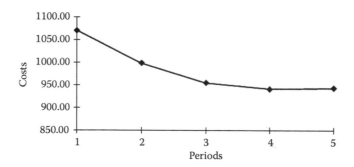

FIGURE 3.4
Evolution of the value of total average cost reduced to present value.

TABLE 3.5

Table for Determining the Lifespan of an Asset

Years	0	1	2	3	4	5	6	7
Acquisition	*1200.00*							
Exploration		*740.00*	*770.00*	*840.00*	*1000.00*	*1200.00*	*1800.00*	*2300.00*
Present value (C_M)		685.19	660.15	666.82	735.03	816.70	1134.31	1342.03
Cession (V_{Cn})		*880.00*	*640.00*	*440.00*	*250.00*	*70.00*	*0.00*	*0.00*
Present value (V_{Cn})		814.81	548.70	349.29	183.76	47.64	0.00	0.00
Exploration + devaluation		1070.37	1311.45	1517.53	1751.27	1969.06	2334.31	2542.03

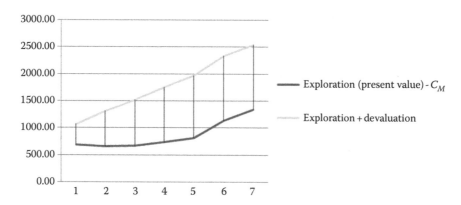

FIGURE 3.5
Analysis of the lifespan.

It can be seen either from Table 3.5 or Figure 3.5 that the replacement interval has four periods. The calculations for the first two periods are illustrated below.

Considering the formula

$$C'_n = \frac{1}{n} \sum_{i=1}^{n} \frac{C_{Mi}}{(1+i_A)^i} = \frac{1}{n} \sum_{i=1}^{n} C'_{Mi} \qquad (3.18)$$

TABLE 3.6

Historic and Predicted Data Evaluated by the Minimization Total Cost Method

Years	−3	−2	−1	0	1	2	3	4	5	6	7
Acquisition				10,200.00							
Acquisition (present value—PV)	16,186.12	12,849.06	11,016.00	10,200.00	9444.44	8097.09	6427.73	4724.57	3215.47	2026.29	1182.32
Exploration cost (EC)	740.00	740.00	740.00		740.00	740.00	750.00	800.00	950.00	1500.00	2000.00
Present value (EC-PV)	932.19	863.14	799.20		685.19	634.43	595.37	588.02	646.55	945.25	1166.98
C'_n	310.73	897.66	2594.52		685.19	659.81	638.33	625.75	629.91	682.47	751.69
Cession (V_{Cn})	13,000.00	12,000.00	10,000.00		9000.00	8500.00	8400.00	8000.00	7000.00	5000.00	4000.00
Cession (V_{Cn}-VP)	16,376.26	13,996.80	10,800.00		8333.33	7287.38	6668.19	5880.24	4764.08	3150.85	2333.96
C''_n	2058.75	1898.40	600.00	10,200.00	1866.67	1456.31	1177.27	1079.94	1087.18	1174.86	1123.72
$C'_n + C''_n$	2369.48	2796.06	3194.52	10,200.00	2551.85	2116.12	1815.60	1705.69	1717.10	1857.33	1875.41
EC-VP accumulated	932.19	1795.32	2594.52	10,200.00	685.19	1319.62	1914.99	2503.01	3149.57	4094.82	5261.80

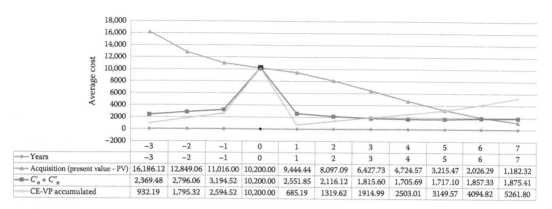

	−3	−2	−1	0	1	2	3	4	5	6	7
Years	−3	−2	−1	0	1	2	3	4	5	6	7
Acquisition (present value - PV)	16,186.12	12,849.06	11,016.00	10,200.00	9,444.44	8,097.09	6,427.73	4,724.57	3,215.47	2,026.29	1,182.32
$C'_n + C''_n$	2,369.48	2,796.06	3,194.52	10,200.00	2,551.85	2,116.12	1,815.60	1,705.69	1,717.10	1,857.33	1,875.41
CE-VP accumulated	932.19	1,795.32	2,594.52	10,200.00	685.19	1319.62	1914.99	2503.01	3149.57	4094.82	5261.80

FIGURE 3.6
Minimization of total of average cost method—reduced to present value.

where:

$$C'_{Mi} = \frac{C_{Mi}}{(1+i_A)^i} \tag{3.19}$$

$$C''_n = \frac{V_A - (V_{Cn}/(1+i_A)^n)}{n} = \frac{V_A - V'_{Cn}}{n} \tag{3.20}$$

where:

$$V'_{Cn} = \frac{V_{Cn}}{(1+i_A)^n}, \text{ it comes to}$$

$$C'_{M1} = \frac{740}{(1+0.08)^1} = 685.19$$

$$C'_1 = \frac{685.19}{1} = 685.19$$

$$Vc'_1 = \frac{880}{(1+0.08)^1} = 814.81$$

$$C''_1 = \frac{1200 - 814.81}{1} = 385.19$$

$$C_1 = 685.19 + 385.19 = 1070.37$$

$$C'_{M2} = \frac{770}{(1+0.08)^2} = 660.15$$

$$C'_2 = \frac{685.19 + 660.15}{2} = 672.67$$

$$Vc'_2 = \frac{640}{(1+0.08)^2} = 548.70$$

TABLE 3.7

Minimization Total of Average Cost Method—Reduced to Present Value, with ROI

Years	−3	−2	−1	0	1	2	3	4	5	6	7
Acquisition				10,200.00							
Acquisition (Present Value—PV)	16,186.12	12,849.06	11,016.00	10,200.00	9444.44	8097.09	6427.73	4724.57	3215.47	2026.29	1182.32
Exploration Cost (EC)	740.00	740.00	740.00		740.00	740.00	750.00	800.00	950.00	1500.00	2000.00
Present Value (EC-PV)	932.19	863.14	799.20		685.19	634.43	595.37	588.02	646.55	945.25	1166.98
C'_n	310.73	897.66	2594.52		685.19	659.81	638.33	625.75	629.91	682.47	751.69
Cession (V_{Cn})	13,000.00	12,000.00	10,000.00		9000.00	8500.00	8400.00	8000.00	7000.00	5000.00	4000.00
Cession (V_{Cn}-VP)	16,376.26	13,996.80	10,800.00		8333.33	7287.38	6668.19	5880.24	4764.08	3150.85	2333.96
C''_n	2058.75	1898.40	600.00	10,200.00	1866.67	1456.31	1177.27	1079.94	1087.18	1174.86	1123.72
$C'_n + C''_n$	2369.48	2796.06	3194.52	10,200.00	2551.85	2116.12	1815.60	1705.69	1717.10	1857.33	1875.41
EC-PV Accumulated	932.19	1795.32	2594.52	10,200.00	685.19	1319.62	1914.99	2503.01	3149.57	4094.82	5261.80
Profit					3000.00	4000.00	5000.00	6000.00	5000.00	5000.00	4000.00
Profit—PV					2777.78	3429.36	3969.16	4410.18	3402.92	3150.85	2333.96
ROI—PV					−7422.22	−3992.87	−23.71	4386.47	7789.39	10,940.24	13,274.20

$$C_2'' = \frac{1200 - 548.70}{2} = 325.65$$

$$C_2 = 672.67 + 325.65 = 998.32$$

The above calculations show a discrepancy in the replacement period that may, at first glance, create a fuzzy space in support of the decision maker. This divergence can be explained from the observation of the graphics, because there is a certain flatness at the minimum point such that, when viewed in conjunction with the respective values, it appears that in the three methods, the differences are irrelevant, particularly in the previous and subsequent points in relation to that.

In the second method, because it does not consider the capitalization and inflation rates, there must be some precautions in relation to its results, namely when we are experiencing an inflationary economy.

Regardless of the results in the second, third, and fourth periods, respectively, the strong flattening in that time interval allows the decision maker a larger time interval to decide the appropriate time to replace the asset. Additionally, and according to what was stated above, the equipment should be replaced by the third or fourth year.

3.8.2 Determination of the Lifespan Replacement

The lifespan of equipment ends when its maintenance costs exceed the maintenance costs plus the capital amortization of an equivalent new equipment. The useful life is usually superior to the economic life.

For the purposes of the calculation of lifespan method, it is necessary to collect the historical cost data of the asset, as shown in Table 3.6 and Figure 3.6. In this example, it turns out that the equipment reaches its useful life after six years.

3.9 Case Study

It was mentioned in previous sections that it is important to evaluate the LCC independently of the time of the life cycle where the asset currently is. This is particularly relevant for

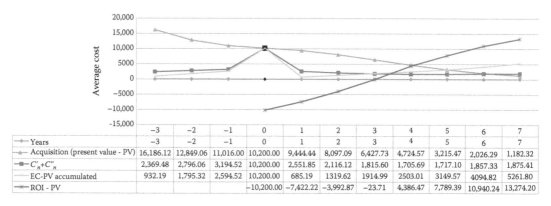

Years	−3	−2	−1	0	1	2	3	4	5	6	7
Years	−3	−2	−1	0	1	2	3	4	5	6	7
Acquisition (present value - PV)	16,186.12	12,849.06	11,016.00	10,200.00	9,444.44	8,097.09	6,427.73	4,724.57	3,215.47	2,026.29	1,182.32
$C_n' + C_n''$	2,369.48	2,796.06	3,194.52	10,200.00	2,551.85	2,116.12	1,815.60	1,705.69	1,717.10	1,857.33	1,875.41
EC-PV accumulated	932.19	1,795.32	2,594.52	10,200.00	685.19	1319.62	1914.99	2503.01	3149.57	4094.82	5261.80
ROI - PV				−10,200.00	−7,422.22	−3,992.87	−23.71	4,386.47	7,789.39	10,940.24	13,274.20

FIGURE 3.7
Minimization of total of average cost method—reduced to present value, with ROI.

TABLE 3.8

Minimization of Total of Average Cost Method—Reduced to Present Value with ROI and Lifespan with Accumulated Cost

Years	−3	−2	−1	0	1	2	3	4	5	6	7
Acquisition				10,200.00							
Acquisition (present value—PV)	16,186.12	12,849.06	11,016.00	10,200.00	9444.44	8097.09	6427.73	4724.57	3215.47	2026.29	1182.32
Exploration cost (EC)	740.00	740.00	740.00		740.00	740.00	750.00	800.00	950.00	1500.00	2000.00
Present value (EC-PV)	932.19	863.14	799.20		685.19	634.43	595.37	588.02	646.55	945.25	1166.98
C'_n	310.73	897.66	2594.52		685.19	659.81	638.33	625.75	629.91	682.47	751.69
Cession (V_{Cn})	13,000.00	12,000.00	10,000.00		9000.00	8500.00	8400.00	8000.00	7000.00	5000.00	4000.00
Cession, present value (V_{Cn}-PV)	16,376.26	13,996.80	10,800.00		8333.33	7287.38	6668.19	5880.24	4764.08	3150.85	2333.96
C''_n	2058.75	1898.40	600.00		1866.67	1456.31	1177.27	1079.94	1087.18	1174.86	1123.72
$C'_n + C''_n$	2369.48	2796.06	3194.52	10,200.00	2551.85	2116.12	1815.60	1705.69	1717.10	1857.33	1875.41
CE-PV accumulated	932.19	1795.32	2594.52	10,200.00	685.19	1319.62	1914.99	2503.01	3149.57	4094.82	5261.80
Profit					3000.00	4000.00	5000.00	6000.00	5000.00	5000.00	4000.00
Profit—PV					2777.78	3429.36	3969.16	4410.18	3402.92	3150.85	2333.96
ROI—VP				−10,200.00	−7422.22	−3992.87	−23.71	4386.47	7789.39	10,940.24	13,274.20
EC-PV Accumulated					685.19	1319.62	1914.99	2503.01	3149.57	4094.82	5261.80

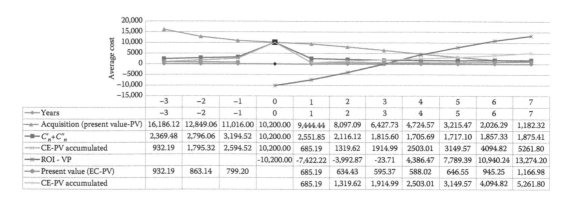

	−3	−2	−1	0	1	2	3	4	5	6	7
Years	−3	−2	−1	0	1	2	3	4	5	6	7
Acquisition (present value-PV)	16,186.12	12,849.06	11,016.00	10,200.00	9,444.44	8,097.09	6,427.73	4,724.57	3,215.47	2,026.29	1,182.32
$C'_n + C''_n$	2,369.48	2,796.06	3,194.52	10,200.00	2,551.85	2,116.12	1,815.60	1,705.69	1,717.10	1,857.33	1,875.41
CE-PV accumulated	932.19	1,795.32	2,594.52	10,200.00	685.19	1319.62	1914.99	2503.01	3149.57	4094.82	5261.80
ROI - VP				−10,200.00	−7,422.22	−3,992.87	−23.71	4,386.47	7,789.39	10,940.24	13,274.20
Present value (EC-PV)	932.19	863.14	799.20		685.19	634.43	595.37	588.02	646.55	945.25	1,166.98
CE-PV accumulated					685.19	1,319.62	1,914.99	2,503.01	3,149.57	4,094.82	5,261.80

FIGURE 3.8
Minimization of total of average cost method—reduced to present value, with ROI and lifespan with accumulated cost.

assets that are not new, where the evaluation must take into account the historic and the future evaluations of its LCC.

Considering the methods presented in the previous sections, a case study considering both situations of the economic cycle and the lifespan will be presented.

The first example uses the minimization total average cost method with reduction to present value. The historical and predicted data are shown in Table 3.7 and the respective graphical behavior in Figure 3.7.

The next example shows the minimization total of average cost method with reduction to present value, but conjugated with the return on investment indicator, also with RPV (Table 3.8 and Figure 3.8).

Similarly to the preceding examples, an example of LCC evaluation independently of the point in the life cycle where the asset is, but based on lifespan, will also be given here.

4

Diagnosis of Maintenance State

4.1 Background

Either the introduction of integrated maintenance management systems in a company and or the need to reorganize its maintenance department, namely to implement new administrative, organizational, and methodological solutions, implies that an adequate diagnosis of the present state of its situation should be made.

To manage this problem, it is particularly useful to use internal audit methods to diagnose the state of maintenance. These methods must be used initially to aid the company's reorganization. After this phase, they should be carried out periodically in order to assess the situation of the maintenance organization and correct deviations from the defined plan.

Diagnosing the maintenance department permits identification of the nonconforming aspects to correct and new solutions to implement. Using this methodology, the reorganization can be implemented consistently and according to the most appropriate model according to the company profile.

The methods to diagnose the maintenance state presented in this chapter have as inputs the data of the current maintenance state and as outputs a set of reports, graphs, and indicators to support the implementation of a reorganization. This will enable more efficient organization and management, as well implementing new approaches for the organization of the maintenance department.

The diagnosis methods also permit permanent monitoring of the maintenance state, as well as the identification, at any time, of any deviations and therefore the definition of corrections to ensure proper consistency between the conceptual model of the organization and the one effectively implemented.

For more on this subject, see works like the following: Farinha (2011) and Raposo et al. (2012).

4.2 Holistic Diagnosis Model

The diagnosis model presented here is called a holistic diagnosis model (HDM) comprising three main components:

1. Basic organization
2. Transversal concepts
3. Management models

The diagnosis method of the maintenance state is based on the use of a sequence of questionnaires, and evaluation of the responses places the situation of the maintenance department in relation to reference landmarks of each subject that was analyzed.

This diagnosis method was designed to be a practical tool for direct use by the responsible of each organization in assessing maintenance management models and also providing systematic measures for their improvement. Additionally, it permits evaluation of the efficiency of the utilization of their equipment and facilities.

The diagnosis of the maintenance state aims to support the implementation of an evolutionary step of this activity, providing it with innovative tools designed to accomplish the best international practices or, in other words, the state of the art. This way is essentially based on two factors that guide the company throughout their application:

1. The first is the introduction of new tools essential to the processing of data for adequate maintenance management of physical assets, with the objective of obtaining an effective management of their life cycle cost.
2. The second is the quality assurance systems through the potential subsequent certification by the appropriate general and specific standards, namely the maintenance ones. To reach this, it is required that the maintenance function better interpret their actions, their operation, and their management.

The developed diagnostic model may be easily implemented through common tools like Excel, or even programmed using a programming language. It is important to have a friendly interface in order to help the user easily understand and interact with the diagnosis model.

The application of the surveys considers their suitability for any company and activity sector, with the concern of not individualizing or limiting the model, which covers three aspects (Table 4.1):

1. Organizational basis
2. Transversal concepts
3. Management models

TABLE 4.1

Steps and Stages of the Questionnaires

Groups	Phases
Organizational basis	1. Technical asset management 2. First-level maintenance 3. Planning and security 4. Databases 5. Maintenance works 6. Spare parts 7. Cost analysis
Transversal concepts	1. RAMS analysis 2. RCM 3. RBM 4. Transversal tools
Management models	1. 5S 2. TPM 3. Lean maintenance

For the organizational basis, seven questionnaires are considered:

1. Technical asset management
2. First-level maintenance
3. Planning and security
4. Databases
5. Maintenance works
6. Spare parts
7. Cost analysis

For the transversal concepts, the following four questionnaires are considered:

1. Reliability, availability, maintainability, and safety (RAMS) analysis
2. RCM
3. RBM
4. Transversal tools

Finally, for management models, three questionnaires are used:

1. 5S
2. TPM
3. Lean maintenance

The final diagnosis is structured from the following fourteen questionnaires:

1. Technical asset management
2. First-level maintenance
3. Planning and security
4. Databases
5. Maintenance works (interventions)
6. Spare parts
7. Cost analysis
8. RAMS analysis
9. RCM
10. RBM
11. Transversal tools
12. 5S
13. TPM
14. Lean maintenance

This method for diagnosing the maintenance state is based on a sequence of the preceding questionnaires, and evaluation of the responses places the position of the maintenance department in reference to landmarks defined in the diagnosis model (Costa et al., 2000; Costa, 2002; Farinha, 2011). The methodology is based on the following phases:

- Data collection—questionnaire responses
- Analysis of the data collected
- Establishment of an improvement action plan

4.3 Questionnaires

The methodology presented here is based on 14 questionnaires, called "diagnosis sheets," to be completed by the company maintenance manager or external consultants. Each questionnaire is accompanied by explanatory text about each question to answer to any questions that may arise. Table 4.2 shows the 14 sheets that constitute the questionnaires that support the diagnosis model, with the respective highest and lowest scores.

In each questionnaire, there are several questions with five possible answers:

1. *Always* (it is always verified)
2. *Almost always* (not always found)
3. *Sometimes* (it is sometimes found)
4. *Almost never* (seldom verified)
5. *Never* (never verified)

According to the kinds of answers, the final score is evaluated using the limits referred to in Table 4.2:

- The maximum score is reached when the response to each question is *Always*, and it is equal to the number of total questions on each questionnaire.

TABLE 4.2

The 14 Stages of the Diagnostic Model with the Respective Highest and Lowest Scores

Stage	Activity	Maximum Score	Minimum Score
1	Management of physical assets	17	9.2
2	First-level maintenance	16	7.2
3	Planning and security	18	8.6
4	Databases	12	5.0
5	Maintenance works (interventions)	16	6.9
6	Spare parts	16	8.2
7	Cost analysis	14	6.8
8	RAMS analysis	18	9.5
9	RCM	14	6.4
10	RBM	14	7.0
11	Transversal tools	12	5.4
12	5S	18	10.8
13	TPM	20	12.0
14	Lean maintenance	21	10.6

- The minimum score depends on the type of questionnaire, according to the column and colors of each one. Each column has an evaluation of 1, 0.7, 0.5, 0.3, and 0. The colors are green, yellow, orange, and red.

Figure 4.1 shows Questionnaire 1, in which the global approach that is repeated in all questionnaires can be seen. Obviously, each one has its own specificities according to its theme, as shown by the first.

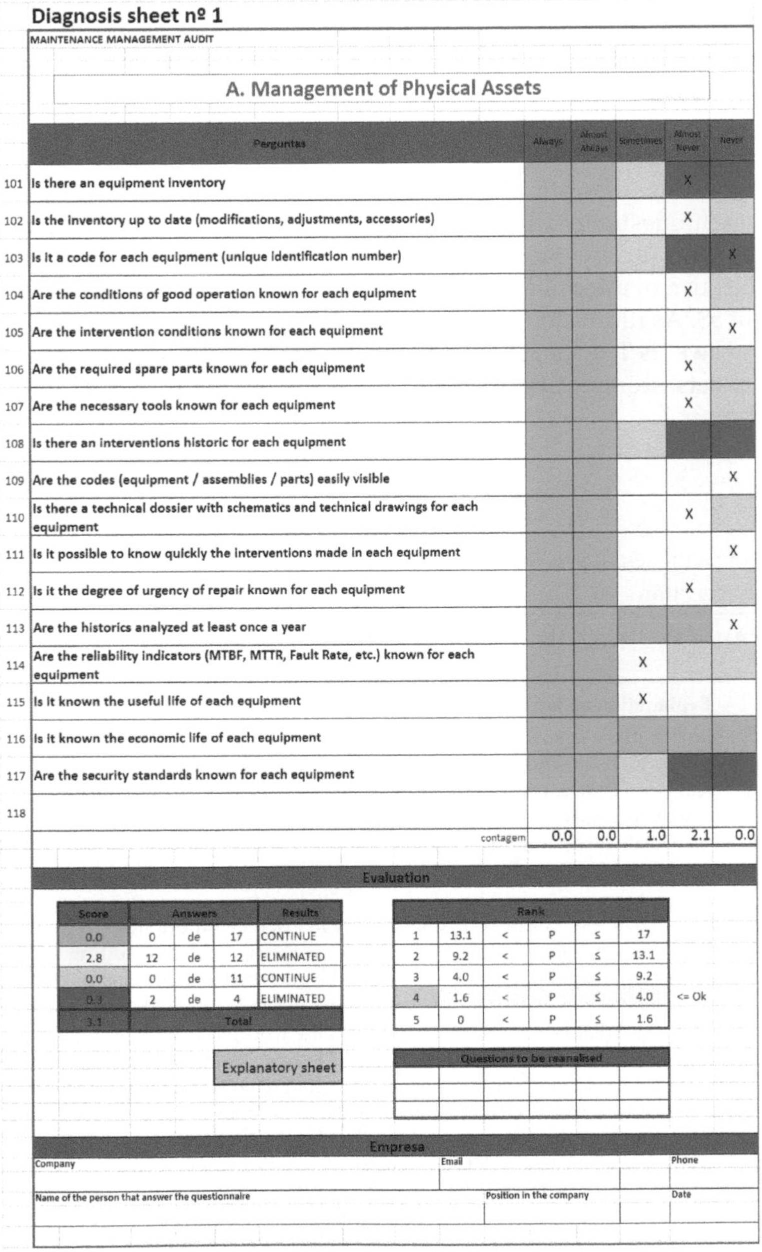

Diagnosis sheet nº 1

MAINTENANCE MANAGEMENT AUDIT

A. Management of Physical Assets

	Perguntas	Always	Almost Always	Sometimes	Almost Never	Never
101	Is there an equipment inventory				X	
102	Is the inventory up to date (modifications, adjustments, accessories)				X	
103	Is it a code for each equipment (unique identification number)					X
104	Are the conditions of good operation known for each equipment				X	
105	Are the intervention conditions known for each equipment					X
106	Are the required spare parts known for each equipment				X	
107	Are the necessary tools known for each equipment				X	
108	Is there an interventions historic for each equipment					
109	Are the codes (equipment / assemblies / parts) easily visible					X
110	Is there a technical dossier with schematics and technical drawings for each equipment				X	
111	Is it possible to know quickly the interventions made in each equipment					X
112	Is it the degree of urgency of repair known for each equipment				X	
113	Are the historics analyzed at least once a year					X
114	Are the reliability indicators (MTBF, MTTR, Fault Rate, etc.) known for each equipment			X		
115	Is it known the useful life of each equipment			X		
116	Is it known the economic life of each equipment					
117	Are the security standards known for each equipment					
118						
	contagem	0.0	0.0	1.0	2.1	0.0

Evaluation

Score	Answers			Results
0.0	0	de	17	CONTINUE
2.8	12	de	12	ELIMINATED
0.0	0	de	11	CONTINUE
0.3	2	de	4	ELIMINATED
3.1			Total	

	Rank				
1	13.1	<	P	≤	17
2	9.2	<	P	≤	13.1
3	4.0	<	P	≤	9.2
4	1.6	<	P	≤	4.0	<= Ok
5	0	<	P	≤	1.6

Explanatory sheet

Questions to be reanalised

Empresa

Company	Email	Phone
Name of the person that answer the questionnaire	Position in the company	Date

FIGURE 4.1
Diagnostic sheet n° 1.

However, it is necessary to consider the questions in the business context. For example, if the company subcontracts all maintenance services, it will not be able to answer questions related to the implementation of new internal works. In this case, it should not mark any response. If fewer than 25% of the questions in each questionnaire are answered, then this is not taken into account when evaluating the final assessment. The questionnaire is considered not to meet the minimum conditions for a sustained evaluation.

4.4 The Explanation Sheets

In order to minimize doubts about the content and improve understanding of the questions in each questionnaire, each one is accompanied by an explanatory sheet that allows the user, question by question, to understand the question and its various choices.

With the objective of exemplifying this, Table 4.3 shows the first question of Diagnostic Sheet 12, about 5S. As illustrated, the example question is given an interpretation showing the choice of answer is True or False and also a suggestion for improvement.

Diagnosis sheets are identified at the top by the stage number (1–14) and the term "Diagnosis sheet n°."

TABLE 4.3

(Questionnaire 12—Question 1) For Each Question, Two Explanations and an Improvement Suggestion

1201	Separate the Equipment, Tools, and Materials: Useful from Useless, Eliminating the Unnecessary Ones (Seiri)	
Explanation of True		**Explanation of False**
The useful are separated from the useless, eliminating unnecessary items. The work begins putting things in order to use only what is really necessary and applicable. So, it is important to have only the necessary items in appropriate quantities, controlled to facilitate operations.		Does not separate the useful from the useless. It does not eliminate the unnecessary.

Improvement suggestion

Separate the useful from the useless, eliminating unnecessary items. It is important to have the necessary items in adequate quantities, controlled to facilitate operations. The work begins putting things in order to use only what is really necessary and applicable. It can also be interpreted with sense of use, storage, organization, and selection. It is essential to know how to separate and classify the useful objects from the unnecessary ones as follows: what is always used is put near the workplace; what is almost always used is put near the workplace; what is occasionally used is put a little away from the work place; what is rarely used, but necessary, is placed separately in a particular place; and what is unnecessary is retired, sold, or eliminated, because it occupies space needed for working objects. Advantages: It reduces the need for and expense of space, stock, storage, transportation, and insurance; it facilitates internal transport, physical arrangement, and production control; it avoids the purchase of materials and components in duplicate and damage to materials or stored products; it increases the productivity of the equipment and people involved; it brings a greater sense of humanization, organization, and economy, less physical fatigue, and greater ease of operation; and it reduces risks of accidental use of these materials by staff. All team members must be able to distinguish the useful from the useless, and what it is really necessary from what is not.

An intermediate zone follows, where there is a grid with the questions and the columns reserved for responses. Each line begins with the indication of the number associated with each statement, consisting of three or four digits. The first represents the number of the questionnaire and the remaining two identify the number of the question and are indexed to the field "Explanatory sheet."

The final answers may have five possible options (Figure 4.1):

1. Always
2. Almost always
3. Sometimes
4. Almost never
5. Never

Finally, the bottom is reserved for the determination:

- Of the obtained score
- Of the resulting rank categories
- Of the achieved elimination criteria

4.5 Organization and Analysis of Information Collected

Depending on the responses to the questionnaire, which interprets the current maintenance state, each diagnostic record reaches a certain score that is classified in a rank with five levels (Figure 4.1):

1. Level 1
2. Level 2
3. Level 3
4. Level 4
5. Level 5

The most positive situation is Level 1, which is synonymous of a very good position in relation to the highest value of the rank. Level 2 translates into a good position. The negative situations are the next levels: Level 3 indicates that there are aspects to be improved in the organization; Level 4 reflects poor performance of the maintenance department, indicating that a broad and deep intervention must be carried out to reorganize it; and Level 5 indicates a bad or nonexistent maintenance organization.

Each questionnaire and stage must achieve a minimum score in order to sustain the position of the next stage. That is, the company cannot adequately ensure the implementation of the issues of a certain stage without the previous stage having reached a certain position that is considered positive. In practice, it is established that the company should reach the

TABLE 4.4

Criteria Importance of Responses in the Position of State Maintenance

Green	**Adequate response** This answer is always desirable.
Yellow	**Inadequate response** Only some answers ought to be this way and the company should improve them.
Orange	**Exceptional response** Few answers should be of this type; although these responses are not qualifiers, the company must improve them as soon as possible.
Red	**Critical response** The company should never have this kind of response, and these should be the first to be reviewed.

threshold of the third category as the minimum sustainability to ensure the implementation of the next stage.

4.6 Elimination Criteria

For every possible response to the questions, a degree of importance is assigned that works as a criterion for elimination, classified in four colors—green, yellow, orange, and red—according to the interpretation given in Table 4.4.

It is considered an elimination criterion if the company does not respond to a question of critical importance or exceeds the maximum number of allowable responses of exceptional importance.

The final evaluation of the questionnaires permits determination of the status of the maintenance department in terms of the organization and management over the equipment and facilities of the organization.

It can be seen in several questionnaires that the columns relating to responses like "always" or "almost always" are the most desirable possibilities of response, so they are always green for the elimination criterion. However, if the answer is negative, in this case, this is not a critical point.

For each question answered with the negative options, that is, "Sometimes," "Almost Never," or "Never," the model automatically prepares a report of weak points (responses obtained in orange areas) or critical points (responses obtained in red zones).

4.7 Elimination Grid

Qualifying grids are no more than colored cells in red, orange, yellow, and green that are part of the grid of each diagnosis questionnaire. The answers can be critical, exceptional, inadequate, and adequate, respectively, allowing one to identify if the company, in each question, is eliminated or not, according to the process described above.

TABLE 4.5

Application of the Elimination Criteria to the Possible Answers

	Always	Almost always	Sometimes	Almost never	Never	
1						The answer to the question is not fundamental to make the maintenance state diagnosis or for the implementation of a new management model. A negative answer of this type of question only contributes to the positioning in a category of poor technical maintenance management.
2						The answer to the question is not fundamental to make the maintenance state diagnosis or for the implementation of a new management model. However, if the company responds "Never," it is desirable to improve this positioning as soon as possible.
3						The answer to the question is not fundamental to make the maintenance state diagnosis or for the implementation of a new management model. However, if the company responds "Never" or "Almost Never," it is desirable to improve this positioning as soon as possible.
4						The answer to the question is not fundamental to make the maintenance state diagnosis or for the implementation of a new management model. However, if the company responds "Never" or "Almost Never," it is desirable to improve this positioning as soon as possible.
5						The answer to the question is important to make the maintenance state diagnosis and for consequent implementation of a new management model. However, a negative positioning of the company on the various questions of this type may put into question an improvement program and should therefore contribute to the elimination of the stage under analysis.
6						The answer to the question is important to make the maintenance state diagnosis and for consequent implementation of a new management model. However, a negative positioning of the company on the various questions of this type may put into question an improvement program and should therefore contribute to the elimination of the stage under analysis.

(Continued)

TABLE 4.5 (*Continued*)

Application of the Elimination Criteria to the Possible Answers

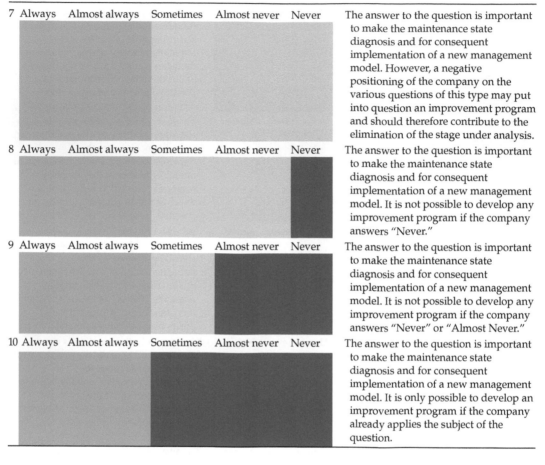

7 Always	Almost always	Sometimes	Almost never	Never	The answer to the question is important to make the maintenance state diagnosis and for consequent implementation of a new management model. However, a negative positioning of the company on the various questions of this type may put into question an improvement program and should therefore contribute to the elimination of the stage under analysis.
8 Always	Almost always	Sometimes	Almost never	Never	The answer to the question is important to make the maintenance state diagnosis and for consequent implementation of a new management model. It is not possible to develop any improvement program if the company answers "Never."
9 Always	Almost always	Sometimes	Almost never	Never	The answer to the question is important to make the maintenance state diagnosis and for consequent implementation of a new management model. It is not possible to develop any improvement program if the company answers "Never" or "Almost Never."
10 Always	Almost always	Sometimes	Almost never	Never	The answer to the question is important to make the maintenance state diagnosis and for consequent implementation of a new management model. It is only possible to develop an improvement program if the company already applies the subject of the question.

The process of assigning colors according to the criteria of elimination results from the importance of how each question contributes to the maintenance state diagnosis and, as a consequence, its implications for maintenance department reorganization.

The score (*R*) obtained by the company results from the following formula:

$$R = \sum R_S + \sum R_{QS} + \sum R_{AV} + \sum R_{QN} \tag{4.1}$$

where:
   R_S = Answer always
   R_{QS} = Answer almost always
   R_{AV} = Answer sometimes
   R_{QN} = Answer almost never

The score gives rise to the category achieved by the company in each stage or questionnaire.

Table 4.5 shows an example of the application of the elimination criteria to the different possibilities of answers.

4.8 Establishment of an Improvement Action Plan

The final stage of diagnosis allows the establishment or proposal of an improvement action plan in the audited company. After the completion and evaluation of the scores of questionnaires, the model produces several positioning graphics called "radar maps" that graphically show the positioning of the company and the most critical questionnaires that must be taken into account to improve the organization.

The diagnostic model automatically produces six radar maps about the company's positioning radar:

1. General position of the maintenance state
2. General position of the maintenance state by each specific subject
3. Position on the organizational base
4. Position on the transversal concepts
5. Position on the transversal concepts and management models
6. Position on the management models

In addition to the radar map positioning, several reports can be printed that help show a characterization of the maintenance state, such the following:

- The position report
- The critical points report
- The fragile points report
- The weak points report

The position report, as its name indicates, has the objective of showing the organizational maintenance state in its various stages. It takes the diagnostic model to make an immediate diagnosis of the general maintenance state, namely through the radar maps. The other reports allow the positions to be obtained according to their designations (critical, fragile, and weak), as well as offering suggestions for improvement for each indicated point. Figure 4.2 shows an example of a radar map.

4.9 Case Study

The case study for a holistic diagnosing model is illustrated based on three passenger transport companies operating in an European country.

According to Table 4.6, it can be seen that the differences among the companies analyzed are evident, as expressed by the results of their maintenance state diagnoses. It can be said that such differences are not only due to their dimensions and structures, but also to the different management models.

Obviously, there is a big difference between the state of the level of maintenance of Company I and Companies II and III. Company III has a better score at the level of most stages.

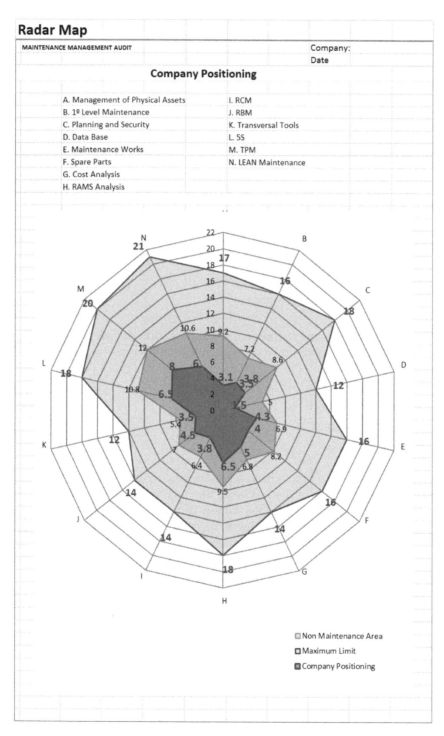

FIGURE 4.2
Radar map.

TABLE 4.6

Comparative Table of the Scores Achieved—Companies I, II, and III

Stages	Activity	Company I Score	Company II Score	Company III Score	Company I Category	Company II Category	Company III Category
1	Physical asset management	9.6	16.1	17	3	1	1
2	First-level maintenance	8.9	15.0	15	2	1	1
3	Planning and security	8.6	15.6	16.5	3	1	1
4	Databases	7.1	11.0	12	2	1	1
5	Maintenance interventions	6.2	15.2	14.5	3	1	1
6	Spare parts	12.1	13.6	14.50	2	1	1
7	Cost analysis	6.8	10.7	13.10	3	1	1
8	RAMS analysis	7.6	15.5	9.5	3	1	3
9	RCM	2.2	3.6	5.6	4	3	3
10	RBM	2.0	6.3	6.8	4	3	3
11	Transversal tools	0.3	2.6	4.8	4	3	3
12	5S	4.2	8.7	10.6	4	3	3
13	TPM	3.1	4.3	10.7	4	4	3
14	Lean maintenance	1.4	4.5	10.3	5	4	3

However, in general, it can be said that Companies I, II, and III should improve their maintenance states. It may be suggested that they ought to introduce new techniques and methods mentioned in the stages to improve their performance and maintenance management.

Figures 4.3a–c illustrate the overall radar maps of each of the diagnosed companies.

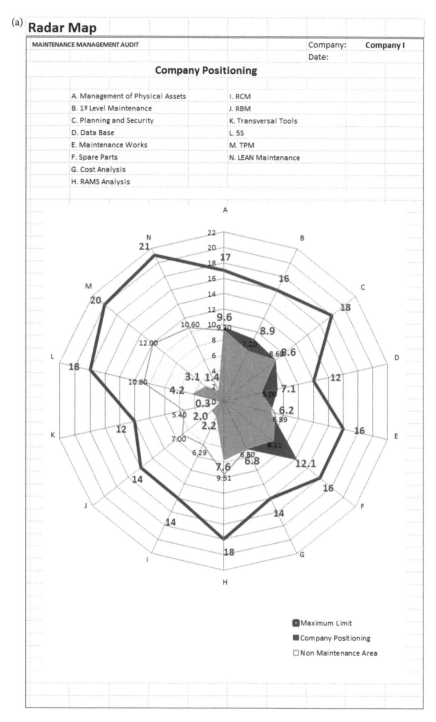

FIGURE 4.3 (*Continued*)
(a) Radar graph—Company I.

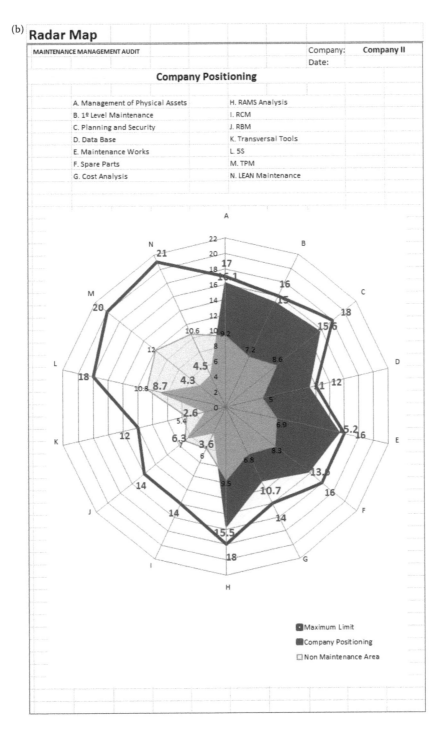

FIGURE 4.3 *(Continued)*
(b) Radar graph—Company II.

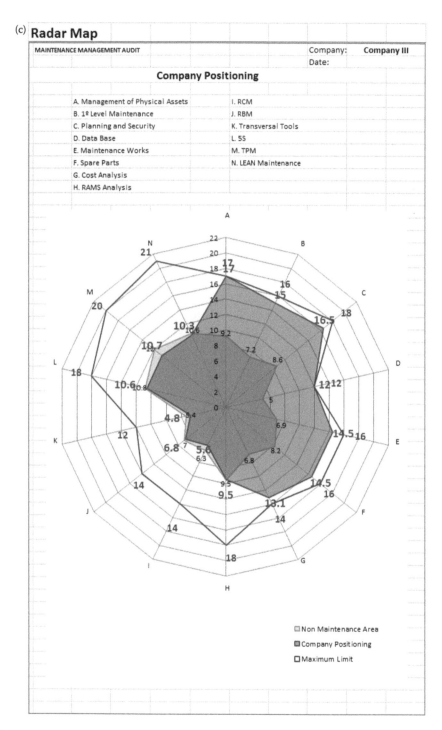

FIGURE 4.3
(c) Radar graph—Company III.

5

Maintenance Management

5.1 Background

This book uses as its main idea the concept of terology, which implies a broad view of the maintenance concept, suggesting that it has connections with several knowledge areas, as discussed in detail in Chapter 2.

This large number of organizational connections implies increased difficulties in asset maintenance management. However, maintenance planning must be one of the most important tasks in any organization. This is the reason this chapter deals with these aspects, namely the last, always indexing the necessary deep developments to other chapters of this book and/or helpful references.

5.2 Maintenance Planning

For decades, maintenance planning was based on calendar scheduling of interventions, which included maintenance actions programmed to perform certain procedures, namely the replacement of components subject to wear and lubrication actions, that were particularly critical for the proper functioning of strategic mechanical equipment.

With the advent of information systems and maintenance engineering as well as techniques from operational research, maintenance planning has explored new frontiers.

Whatever the control variables of planned maintenance, its planning has to take place on a specific date, so this is the challenge that any maintenance planning system has to face.

The next section discusses some maintenance planning techniques that allow projection of the corresponding real-time calendar interventions either for periodic or aperiodic intervals (Farinha, 1994, 1997).

5.2.1 Scheduled Maintenance

The most elementary way of doing maintenance planning is to schedule interventions at regular intervals in time, for example, weeks, months, quarters, or other intervals, according to the specifications of each particular piece of equipment. In this way, using only one calendar, the interventions can be scheduled, and this type of maintenance is called systematic, scheduled maintenance.

The way to plan systematic calendar interventions prior to the advent of information systems was through tables. Nowadays, this approach continues to be very useful, as illustrated in Figure 5.1.

Weeks/ Physical assets	1	2	3	4	5	6	7	...	...	51	52
Physical asset 1											
Physical asset 2											
Physical asset 3											
Physical asset 4											
...											
...											
...											
...											
...											
Physical asset n-2											
Physical asset n-1											
Physical asset n											

FIGURE 5.1
Scheduled maintenance map—calendar.

Although all interventions have to be reported using a calendar date, the most significant part of systematic maintenance is programmed from variables other than a calendar, namely hours of functioning time, number of manufactured parts, and kilometers traveled, among others. From this perspective, it is important to define the various types of systematic interventions and the algorithms that permit mapping the scheduled variable onto the calendar intervention time.

5.2.2 Planned Maintenance through Control Variables

As mentioned in the previous section, the systematic control variables of planned interventions are not, in most situations, based on calendar variables, but on other types of control variables, which must be mapped onto a time calendar.

The easiest way to plan the next intervention is by evaluating the next value for the control variable (CV) through the present value associated with a constant increase. For example, if a given maintenance object (MO) has interventions at intervals of 5000 hours, this will be the increment that must be associated with the last measured value of the CV to determine the next value of this variable, which can be represented by the following formula:

$$x_{t+1} = x_t + I_{nc} \tag{5.1}$$

where x_{t+1} is the next value of the control variable.

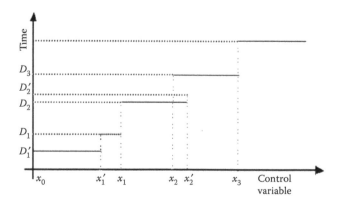

FIGURE 5.2
Systematic planning with constant increment.

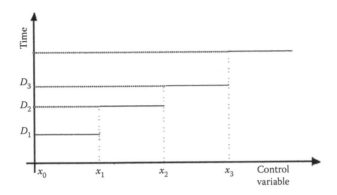

FIGURE 5.3
Systematic planning with baseline increment.

Figure 5.2 shows the behavior of scheduled planning through a constant increment.

However, even if the planning is constant, according to some CVs, many times it is not possible to perform the maintenance intervention at the right time. This can be anticipated or delayed, but the next intervention is always done at the next value of the CV. For example, if an intervention is to be done each 15,000 increment values (IVs), the current value of CV is 32,000 units, and the intervention ought to be done at 30,000, this corresponds to a delay of 2000 units. However, if the scheduled plan is through a baseline increment, then the next intervention in this example is at 45,000. The intervention after that is at 60,000, and the same goes for the next ones—this is the baseline increment. Figure 5.3 shows this case of scheduled maintenance.

5.2.3 Condition Monitoring

Condition monitoring maintenance can be described as preventive maintenance based on the monitoring of the functioning of an asset, namely the most significant variables of its "health."

Condition monitoring, or condition maintenance, which can also be predictive maintenance, is an approach to maintenance planning that is currently widespread in a wide range of industries, such as aeronautics, steel plants, oil refineries, ships, power stations, and so on, and is supported by a set of technologies, namely those developed in the last decades, such as those related to the use of sensors and the monitoring of the performance of assets, which have allowed very significant gains in the reliability and performance of the installations and equipment.

The modern techniques of condition monitoring maintenance of physical assets involve the use of several methods in which the analysis of vibrations and lubricants are of particular importance. These variables are often complemented with information concerning other individual variables, such as pressure and/or operating temperature, among others. However, it should be noted that there are specific techniques that are very relevant to the monitoring of the condition of equipment and are based on the control of variables of the production process.

Among the various technologies associated with condition monitoring, the following are particularly relevant:

- Vibration measurement and analysis
- Infrared thermography

- Oil analysis and tribology
- Acoustic emission
- Ultrasonic measurement
- Motor current and voltage analysis
- Thermic motor effluents
- Structural health monitoring

Predictive maintenance that attempts to detect the onset of a degradation mechanism with the goal of correcting that degradation prior to significant deterioration in the component or equipment is associated with condition monitoring. The diagnostic capabilities of predictive maintenance technologies have increased according to innovations in sensor technologies, namely in size reduction, cost, and the Internet of Things.

Nowadays, the IoT is intrinsically connected to condition monitoring, including prediction. The IoT is the outcome of technology advances in several areas, namely the following:

1. Connecting devices and sensors and providing standardized ways to talk in the world of sensors
2. Ubiquitous data networks, where the companies can build, at low cost, data networks with widespread coverage
3. The rise of the cloud and the shift from enterprise to software as a service (SaaS) platforms, namely the open ones
4. Big data technology, which has the ability to process large amounts of data in a standardized way

This means that each "thing" can be connected and communicate its status back to software platforms, namely to the EAM/CMMS. Cloud-based software platforms built on the latest advances in big data technology can swiftly process this information and offer insights as a direct prerequisite for predictive maintenance.

This subject is discussed in detail in Chapter 12.

5.3 Maintenance Control

Maintenance control is done based on several areas where maintenance activity acts. However, one of the main transversal aspects supporting it is working orders. One of the most important times in its control is the moment of its closing. Figure 5.4 shows a circuit of control of an intervention, including this last step.

Maintenance control is accomplished through many aspects, as is referred to throughout this book. However, KPIs are the most important guide to accompany the performance of maintenance activity. The ratios can be compared to a cockpit, because when this doesn't work well, the car or plane, for example, is at risk. The same situation happens with companies.

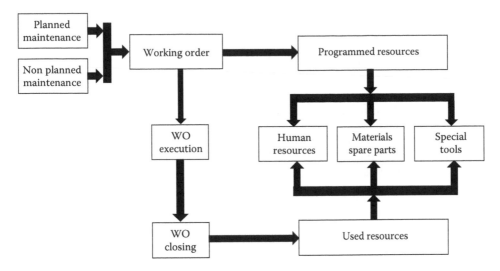

FIGURE 5.4
Circuit of control of an intervention through a WO.

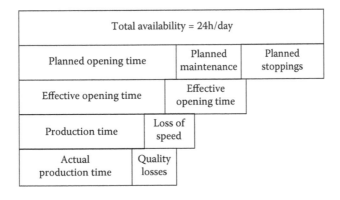

FIGURE 5.5
OEE evaluation.

EN 015341:2009 (*Maintenance Key Performance Indicators*) is a norm that can be used to implement a "cockpit chart" with the ratios that best fit the company maintenance activity, usually classified into three categories: technical, economical, and organizational.

When a company creates its KPI chart, it must take into account its internal activity, but also its competitors, which implies creating indicators based on a pragmatic strengths, weaknesses, opportunities, threats (SWOT) analysis, where the best homologous companies are rigorously compared—the benchmark concept must be always present.

One important indicator not objectively referred to in the KPI norm is the overall equipment efficiency (OEE), which measures the efficiency, or, in other words, the operational availability, with which a production physical asset is used. The OEE allows identifying the factors that negatively influence the efficiency of a production asset.

For industrial companies, this KPI is very relevant, being used in the following slogan: World-class companies have OEEs above 85%. One calculation method is shown in Figure 5.5.

Equations 5.2 and 5.3 show two ways to calculate the OEE. Figure 5.5 shows the variables that contribute to the OEE evaluation.

$$OEE = \frac{Actual\_Production\_Time}{Planned\_Opening\_Time} * 100 \qquad (5.2)$$

$$OEE = \frac{Number\_of\_good\_products*Technical\_cycle\_time}{(Total\_Availability - Planned\_Maintenance - Planned\_Stoppings)} * 100 \qquad (5.3)$$

World-class companies ought to have an OEE greater than 85% and, if they fulfill this requisite, they may be nominated for the TPM Award.

Another way to evaluate the OEE ratio is the following:

$$OEE = Availability*Performance*Quality \qquad (5.4)$$

The variables involved in Equation 5.4 correspond to the following:

- *Availability* is the measurement of stop losses, or, in other words, the real capacity for production.
- *Performance* is the measurement of losses by the production rate variation.
- *Quality* is the measurement of losses due to defects in products.

The minimum value for OEE in order for the company to be considered world-class is given by:

- 90% for *Availability*
- 95% for *Performance*
- 99% for *Quality*

The following example presents a way to calculate the OEE (Table 5.1).

Planned production time = (Shift duration − Pauses) = (480 − 2∗30) = 420 mn
Operating time = (Planned production time − Stop time) = (420 − 37) = 383 mn
Good pieces = (Total pieces produced − Rejected pieces) = (19,172 − 423) = 18,749 pieces
Availability = (Functioning time/Planned production time) = (383 mn/420 mn) = 0.912 = 91.2%
Performance = ((Total of pieces/Functioning time)/Ideal processing rate) = ((19,172/383)/60) = 0.834 = 83.4%
Quality = (Good pieces/Total of pieces) = 18,749/19,172 = 0.978 = 97.8%
OEE = Availability∗Performance∗Quality = 0.912∗0.834∗0.978 = 0.7438 = 74.38%

TABLE 5.1

Example of Production Values for OEE Evaluation

Item	Data
Duration of shift	2 × 4 hours = 480 minutes
Short breaks	2 × 15 minutes = 30 minutes
Food breaks	1 × 30 minutes = 30 minutes
Stop time	37 minutes
Ideal processing rate	60 pieces per minute
Total of pieces produced	19,172 pieces
Rejected pieces	423 pieces

5.4 Maintenance Resources

The maintenance resources are the following:

- Human resources (HR)
- Material resources, namely the spare parts
- Tools
- Financial

5.5 Maintenance Budget

The maintenance budget ought to be made according to the strategic asset management plan. Annually, the maintenance budget has to be designed according to the company's activity plan, and this implies its compliance with the aspects referred to next:

- Resource costs:
  - Human resources
  - Materials (spare parts)
  - Tools
  - Outsourcing
- Planned maintenance costs
- Unplanned maintenance costs
- Structural costs
- Profits:
  - Availability versus maintenance budget
  - Production versus maintenance budget
- Amortization and reintegration

Additionally, it is important to point out that it is more and more necessary to change the philosophy of the maintenance budget from the perspective of maintenance costs to the perspective of increasing the company's profits. This subject is discussed in Chapter 3, namely Section 3.4.

It is usual to say that maintenance is a cost: that is an idea constructed over several decades. However, it is necessary to quickly change the financial perspective on maintenance. In fact, if the assets are necessary to production, that is, to create profit, that implies they have adequate maintenance to have the required availability to production. This implies that maintenance must have a budget, controlled by KPIs that are well defined and indexed to their LCCs. This means that, from the beginning of an asset's life cycle, it is necessary to rigorously conjugate the asset's production with its life cycle cost or, in other words, with its potential availability for production over its lifetime.

The last aspects are almost always omitted from a company's planning and budgeting, which implies the idea that the equipment can produce without maintenance. Then, when a problem occurs, maintenance activity is looked at as an expensive and boring problem.

One problem that occurs in almost every situation is the question of what the most rational KPIs are that ought to be used for each type of asset. To answer this, it is desirable to make a good benchmark, both at the national and international level, to make it possible to evaluate the evolution of internal performance and its comparison to the best references.

For many assets, there are international KPIs that relate their acquisition cost to their maintenance value. However, the maintenance costs vary over their lives and, obviously, their availability and, as a consequence, their production capability. Other important aspects to take into account are those related to environmental conditions and also production capacity. For example, the management requisites are different for an asset that works 8 hours a day and a similar asset that works 24 hours a day.

The conclusion of the preceding remarks is that the KPIs related to asset profits versus maintenance costs vary over the asset's economic life.

5.6 The Strategic Asset Management Plan

What is usually called a strategic plan is, in practice, a development plan, which includes strategy, tactics, and a short-term approach to implementing actions. The same applies to the strategic asset management plan, but focused on—as its designation implies—asset management.

5.6.1 The Asset Development Plan

The asset development plan (ADP) of an organization implies that it has a clear definition of its mission, vision, and politics:

- A mission statement quickly defines a company profile and its purpose. It is an eloquent and concise phrase that should be full of meaning and impact. The mission statement should cover the goals for the stakeholders, namely customers, employees, and suppliers.

- A vision statement ought to clearly and concisely communicate the business's overall goals and serve as a tool for strategic decision-making across the company. A vision statement can be simple and intuitive, like a single phrase or a short paragraph. Effective vision statements define the core ideals that give the company direction and represent its goals.

- The politics statement is a document that expresses the directives of the top management with an emphasis on quality. Quality policy management is a strategic item. In the ambit of Strategic Asset Management Plan (SAMP), referred to in ISO 55001, ISO 9001 may help a lot. For example, Section 5.1 of the former requires a written, well-defined quality policy that must be communicated and understood by the whole organization. Section 5.2 also sets out some of the requirements for quality policies.

Only organizations with a long-term perspective are able to make a good ADP, which can be designed using three levels:

- *First level—Long-term planning*: This is called the strategic plan, because it contains a long-term vision with global objectives. This is what, in the ambit of this book, is called the strategic asset management plan:

- Horizon of three or more years.
- Strategy aligned with the company's management and policy.
- Allows programming the workload in the defined horizon.
- Allows programming the necessary resources.
- Involves defining a maintenance policy.
- *Second level—Medium-term planning*: This is called the tactical plan, because it defines ways to reach the macro objectives defined by the strategic plan:
    - Medium-term horizon, but with definitions of execution tools.
    - From the planned outputs, a load plan is established.
    - Allows planning of maintenance interventions and the necessary resources for them.
- *3rd Level—Short-term planning*: This is called the operational plan, or activities plan, usually with a schedule for a year. This plan defines objectively what the milestones and deliverables are for the company within a year, usually an economic year:
    - Horizon of one year—Activities plan.
    - Maintenance planning in the economic year.
    - Manages and launches maintenance actions.
    - Plans all the equipment and tools necessary for all maintenance interventions.

Additionally, the asset development plan and *Hoshin Kanri* ought to be referenced and compared. This is a management and control system for the organization focused on strategy.

Hoshin Kanri brings significant improvement in organizational performance by aligning the activities of all sectors within the organization with the strategic goals.

Hoshin Kanri was developed in the 1960s by Japanese companies as a method to manage the achievement of strategic objectives throughout the organization's functional structure. Its principle is that each component of the organization must incorporate within its routine of action the corresponding contribution to the overall objectives of the company. They are two ways or, in other words, two cultures to reach similar objectives.

5.6.2 The Strategic Asset Management Plan and ISO 5500X

The strategic asset management plan is referred to in several clauses of the ISO 5500X standards, namely in the following: ISO 55000 at Clauses 3.3.2 and 3.3.3; ISO 55001 at Clauses 4.1, 4.3, 4.4, 5.1, 5.3, 6.2.1, and 6.2.2; and ISO 55002 at Clauses 4.1.1.1, 4.1.1.2, 4.1.1.3, 4.2.4, 4.3, 4.4, 5.2, 6.2.1.1, 6.2.2.1, 7.2.2, and 8.3.3. These links between ISO 5500X standards and the SAMP demonstrate the relevance of this document to the company's certification.

ISO 55000 in Clause 3.3.2 defines a strategic asset management plan as follows: "Documented information that specifies how organizational objectives are to be converted into asset management objectives, the approach for developing asset management plans, and the role of the asset management system in supporting the achievement of asset management objectives."

Hastings (2015) proposes an outline for the contents of the SAMP.

5.6.3 Implementing a Strategic Asset Management Plan

To implement a SAMP, it is almost mandatory for a company to have a global development plan that covers all aspects of the organization—only with this assumption is possible to have a robust SAMP.

To start the process, it is fundamental to know all the physical assets and preferably to classify them through ABC analysis or another method. This objective is easily reached if the company has a computerized maintenance management system or enterprise asset management system correctly uploaded with the data of the physical assets, like the one referred to in Chapter 7.

It is fundamental to evaluate the following aspects:

- The situation of the life cycle cost for existing assets or the foreseeable LCCs of new ones
- The production necessities, namely in a long-term vision
- Any new foreseen acquisitions
- Asset operation, real and foreseen
- Asset maintenance needs
- Real or foreseeable withdrawal
- Amortizations, reintegration, and funding for replacements

The SAMP ought to be reviewed annually in order to correct deviations and incorporate new data according to the reality of the latest period. For this, beyond the regular evaluation of the eventual deviations of the SAMP's implementation, it is mandatory to periodically (usually each year) compare eventual deviations between the defined KPIs and the target values and make the necessary corrections in order to implement the SAMP, always keeping it updated according to the necessary corrections referred to previously.

A powerful tool that can be added to strengthen the SAMP is the balanced scorecard (BSC). This tool can be used by companies to keep track of the execution of activities by the staff and to monitor the results from these actions. The BSC is usually used in two forms:

1. As a strategic management tool, as originally defined by the authors
2. As individual scorecards containing evaluations to manage the company's performance

5.7 Case Study

There are many ways to create a maintenance budget and control its execution. Because of this, the following tables and figures show a practical approach to a real maintenance budget. The case study refers to a passenger urban transport company in a European country, with a fleet of 100 buses and a reserve fleet of 20%.

Table 5.2 presents the evolution of the budget execution through September and the prediction of budget execution until the end of the year 2016. Figure 5.6 graphically shows

TABLE 5.2

Evolution and Prediction of Budget Execution in the Year 2016

Description	Initial forecasting	Execution until August	2016												Forecasting from September to December 2016	Total 2016
			January	February	March	April	May	June	July	August	September	October	November	December		
Maintenance	838,805.51	683,297.18	52,426.21	93,961.47	119,167.84	53,842.17	97,606.02	90,555.84	94,082.66	81,654.97	70,449.65	80,449.65	70,449.65	80,449.65	301,798.59	985,095.77
Buildings and facilities	21,141.40	30,044.76	435.07	681.35	1210.37	491.61	14,522.26	1216.29	1647.33	9840.48	3755.60	3755.60	3755.60	3755.60	15,022.38	45,067.14
Basic equipment	714,554.80	556,299.13	42,300.52	86,545.95	86,692.94	51,442.94	80,609.83	79,112.83	67,295.01	62,299.11	54,537.39	64,537.39	54,537.39	64,537.39	238,149.57	794,448.70
Transport equipment	14,321.72	4377.08	953.35	705.53	351.36	624.05	144.40	992.13	165.53	440.73	547.14	547.14	547.14	547.14	2188.54	6565.62
Tools	1816.31	2616.64	54.16	202.62	640.87	558.79	237.59	162.14	154.89	605.58	327.08	327.08	327.08	327.08	1308.32	3924.96
Administrative equipment	5606.73	10,352.68	866.40	3655.76	1365.76	814.26	539.22	911.52	1057.96	1141.80	1294.09	1294.09	1294.09	1294.09	5176.34	15,529.02
Other immobilizations	66,064.64	32,561.90	7816.71	2170.26	7976.27	2827.39	1852.72	5846.46	1466.34	2605.75	4070.24	4070.24	4070.24	4070.24	16,280.95	48,842.85

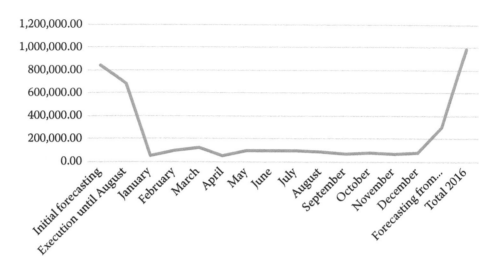

FIGURE 5.6
Evolution and prediction of maintenance budget execution in the year 2016.

the evolution of and prediction for maintenance budget execution in the year 2016, whose absolute values and the others related to it can be analyzed in detail in Table 5.2.

The analysis of the budget execution ought to be made not only globally but in detail, namely using its most relevant aspects, according to the specificity of each company. In the present case study, because the main equipment for its business is the bus, the analysis of the fleet bus maintenance costs analyzed by homogeneous groups (Figure 5.7) was chosen as an example.

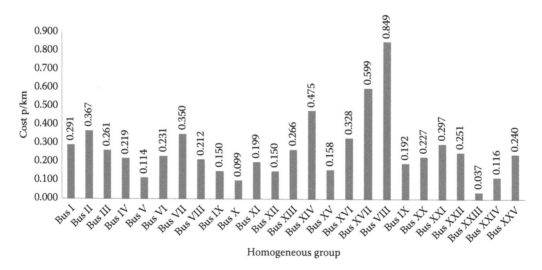

FIGURE 5.7
Fleet bus maintenance costs analyzed by homogeneous groups—year 2016.

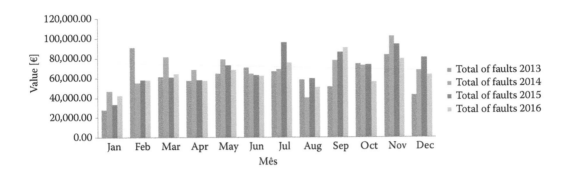

FIGURE 5.8
Cost of faults between 2013 and 2016.

TABLE 5.3

Table of Total Costs—2016

Total faults	1,248,220.08
Total lubrications	29,239.31
Total planned interventions	438,546.21
Total mandatory inspections	304,711.51
Total	2,020,717.11

Finally, the cost of faults between 2013 and 2016 is analyzed, with the objective of evaluating the budget execution and the quality of the maintenance practiced by the company (Figure 5.8).

Table 5.3 discretizes some costs related to the maintenance activity.

6

Maintenance Resources

6.1 Background

Maintenance resources, as referred to previously in Chapter 5, are the following: human resources; material resources, namely spare parts; and tools.

This chapter deals with the design of the main maintenance resources that are fundamental to maintenance management, as discussed in the previous chapter.

6.2 Human Resources

Human resources are the most important maintenance resource, which means special attention should be given to their dimensioning and management.

Human resources have a high impact on and sensitivity to the organization's sustainability. This resource is also conditioned to the market trends and management models that prevail at a given time. From this perspective, several approaches can be designed, from ones that consider that organizations must have all the human resources they need to those that consider that all maintenance services must be outsourced.

Given this diversity of approaches, first of all, it is necessary to say that organizations must know and define the labor hours they need and, from these, decide the best management policy for each specific situation.

From the various variables that must be taken into account about human resources, the following can be highlighted:

- Technicians with adequate training to maintain the existing physical assets
- The life cycle of installed assets and forecast of their evolution with the consequent need of technician retraining
- The diversity of installed technologies
- The availability of qualified technicians in the geographical region where the organization is located
- The geographical dispersion of the organization
- The technicians' average age

- The legal contract constraints
- The market stability for the products or services that the organization provides

According to Bershad (1991), a simplistic but pragmatic way to determine the human resources required is through finding the total number of hours necessary to do the work required, with the constraints that will be analyzed next.

Considering a 35-hour work week and a year with 52 weeks, we will have a total of 1820 hours per year:

$$Total\ working\ time = 35\ hours/week * 52\ weeks/year = 1820\ hours/year$$

This is based on the number of work hours that determine the remuneration of the employee.

Now, nonproductive times will be considered, in particular two daily intervals of 15 minutes for coffee and another 15 minutes for personal hygiene at the end of the work day, which gives the following work time reductions:

$$Total\ nonproductive\ time\ per\ day = 2 * 15\ mn/day + 15\ mn/day = 45\ mn/day$$

This yields:

$$Total\ nonproductive\ time\ per\ year = 45\ mn/day * 5days/week * 52\ weeks/year = 11700\ mn/year$$

The same in hours is:

$$Total\ nonproductive\ time\ per\ year = 11700\ mn \div 60\ mn = 195\ hours/year$$

Then, the percentage of nonproductive time is given by:

$$Total\ nonproductive\ time\ per\ year = 195\ hours/year \div 1820\ hours/year \cong 10.7\%/year$$

If 22 days of vacation per year is assumed and, on average, 10 play days and 4 days of sick leave per year, the productivity time is given by:

$$Nonproductive\ time\ per\ year = 22 + 10 = 4 = 36\ days/year$$

$$\Leftrightarrow$$

$$Nonproductive\ hours\ per\ year = 36\ days/year * 7\ hours/day = 252\ hours/year$$

$$\Leftrightarrow$$

$$Nonproductive\ hours\ per\ year = 252\ hours/year \div 1820\ hours/year \cong 13.8\%$$

The previous calculations do not include some nonproductive times, which are attributed a percentage between 14 and 15% of total gross time (1820 hours):

- Routes between workplaces and workshops
- Requisitions of materials and tools
- WO reading, phone calls, written reports, and oral communications
- Several meetings
- Personal time (toilet, smoking, relaxing, etc.)

In view of the above values, the total unproductive time will be:

$$Total\ nonproductive\ time = 10.7\% + 13.8\% + 14\% = 38.5\% \rightarrow 40\%$$

From the preceding calculations, this will come to a total productive time of about 60%, so about two workers are needed to satisfy a total time of 1820 hours/year.

Now, it is important to assess how the previous approach can be combined with the specific needs of each maintenance department. The calculus of human resources to meet all the needs of maintenance interventions can be made from a planned maintenance requirements framework, as exemplified in Table 6.1.

It is important to emphasize that it is necessary to have a CMMS that is working well with the physical asset dossier and, in particular, with maintenance planning. With these data, the working hours can be added by specialty; then, taking into account the limitations referred to above, the number of human resources the company needs can be evaluated. Obviously, it is necessary to add the working hours, by specialty, of nonplanned working orders, which can be evaluated from the WO history.

Based on these values, there is a basis for calculating the number of human resources required to maintain the physical assets.

For the determination of total human resources, it is necessary to divide the total number of the required human resource total hours/year per specialty by the total number of hours/year of technicians and multiply this by its productivity coefficient. Based on this, the number of technicians, by specialty, needed to fulfill the necessities illustrated in Table 6.1 are presented in Table 6.2.

For the calculation of the number of technicians by specialty, 1820 hours/year with a productivity of 60% are considered. For example, for the electricity specialty:

$$Electricity:\ 4075\ hours/year \div (1820\ hours/technician * 60\%) \cong 4\ technicians$$

Additionally, it is also important to evaluate the technological obsolescence of the physical assets, because it may influence the outsourcing option. In fact, if there is a large number of assets with fast technological obsolescence, the pros and cons ought to be evaluated between contracting internal human resources and outsourcing maintenance services for those assets.

A decision between the preceding options because one model or another is in fashion, or because others are doing the same, should never be made. All decisions, these in particular, must be strongly supported by quantitative analysis within a time frame that permits recovering investments and generating enough added value to justify the option.

TABLE 6.1

Map of Total of Hours Needed by Speciality for Planned Maintenance

Physical Asset	Time by Specialty and Per Year							
Designation	Automation	Electricity	Mechanical	Locksmith	Carpentry	Construction	Garden	Security
Steam generation		100	650					
Medical gases		50	250					
Centralized vacuum		300	200					
Air conditioning		575	100					
Detecting and extinguishing fires		200	100					
Elevators		500	450					
Electric network		700						
Buildings						8300		
Metal furniture				2500				
Wood furniture					5200			
Gardens							7600	
General equipment		1650	20	40				
Diagnostic equipment	7860							820
Treatment equipment	4300							50
Laboratory equipment	3200							20
Totals	**15360**	**4075**	**1770**	**2540**	**5200**	**8300**	**7600**	**890**

TABLE 6.2

Map of Total Human Resources by Speciality for Planned Maintenance

Physical Asset Designation	Time by Specialty and per Year							
	Automation	**Electricity**	**Mechanical**	**Locksmith**	**Carpentry**	**Construction**	**Garden**	**Security**
Total hours	15360	4075	1770	2540	5200	8300	7600	890
Total human resources	14	4	2	2	5	8	7	1

6.3 Spare Parts

Maintenance interventions need materials, usually classified as spare parts. Therefore, it is necessary to know a set of data about them to conduct their proper management. However, the main spare parts data are similar to any other stock material. Figure 6.1 summarizes some essential data of a spare part (Farinha, 1997).

```
                          SPARE PART SHEET


  CODE: _____ - _____ _____

  LOCATION IN WAREHOUSE: _____          TECHNICAL DOCUMENTS: _____

  BRAND: ____ - _____

  UNIT TYPE: _____                       CONVERSION FACTOR: _____

  STOCK:  MINIMUM: _____   MAXIMUM: _____    EXISTING: _____

  LAST PURCHASE: __/ __/ ____                LAST EXIT: __/ __/ ____
  PRICE OF LAST PURCHASE: _____          EXIT PRICE: _____


  ORDERING DOCUMENT: _____                 QUANTITY ORDERED: _____

  ORDER DATE : __/ __/ ____                  TIME TO DELIVER: _____ DAYS

  SUPPLIER: _____ - _____
                        ALTERNATIVE SUPPLIERS

     CODE          SUPPLIER NAME              TIME TO DELIVER
```

FIGURE 6.1
Spare parts sheet.

It is based on these data that spare parts management is done, but, in this case, each specific spare part can be linked to a specific physical asset. Additionally, they are linked to each working order where they are used. From these connections, when included within a CMMS, comes a strategic contribution for a good maintenance management and control.

The management of spare parts is extremely complex due to the diversity of situations that exist in the maintenance activity, added to the costs associated with it.

About this last point, minimizing the costs associated with the stock of spare parts will theoretically be achieved through a Just-In-Time (JIT) approach. However, in the maintenance field, it is almost impossible to reach this situation because of the security of spare parts. In fact, maintenance activity must ensure the maximum availability over the entire asset life cycle, which implies always taking into account planned and, in particular, nonplanned interventions.

This management involves proper planning of acquisitions, which can use either the traditional inventory management techniques or algorithms based on time series, such as those covered in this book, or still others, including neural networks.

Based on the most traditional stock approach, there are two main variables, quantity and time, which implies four combinations of situations, as can be seen in Table 6.3.

The first situation, a supply program, is the easiest one to manage, because the time intervals between acquisitions and the quantities ordered are both constant. The algorithm is described next.

The first step is to identify the main costs involved in stock management, which are the following:

- The administrative cost (C_{at})
- The cost of ordered materials (C_m)
- The cost of materials ownership (C_p)

$$Total\ cost(CT) = Administrative\ cost(C_{at}) + Materials\ cost(C_m)$$
$$+ Ownership\ cost(C_p)$$
$$\Leftrightarrow \qquad\qquad\qquad (6.1)$$
$$CT = C_{at} + C_m + C_p$$

These costs depend on the following variables:

- K—Expected annual consumption (in number)
- Q—Recommended amount for each supply
- N—Annual number of orders
- P_u—Unit price of materials
- i—Possession rate applied to the average annual stock

TABLE 6.3

Methodologies Applicable to Materials Acquisition

Model Type	Quantity to Order	Time between Delivery Orders
Supply program	Periodic	Periodic
Order point	Periodic	Variable
Procurement plan	Variable	Periodic
Security parts	Variable	Variable

- C_a—Administrative cost of order acquisition
- C—Cost of material resale

In order to help with the calculation of the value of some of the previous variables, a detailed analysis will be done next. Starting with the acquisition cost procedure (C_a), this varies according to the number of purchase orders to be issued to each vendor (single order or grouped orders).

These costs are usually subdivided for the following departments:

- Acquisition (provision)
- Stock management
- Reception (quality control)
- Warehouse
- Accounting

Over the year, the total acquisition cost is equal to the unit acquisition cost multiplied by the number of orders:

$$C_{at} = C_a * N \qquad (6.2)$$

$$C_{at} = C_a * \frac{K}{Q} \qquad (6.3)$$

The cost of stock material possession is equal to the *annual possession rate* multiplied by the amount of *property stock*.

The possession rate i is a function of:

- Cost of immobilized capital
- Cost of stocks (space, fees, taxes, insurance)

In each order, a provision of Q quantity of materials is made, as follows:

$$Q = \frac{K}{N} \qquad (6.4)$$

As a linear consumption of materials over time, the evolution of stock quantities can be represented by the graph in Figure 6.2.

According to Figure 6.2, the average stock is equal to $Q/2$ and $[0, T_1] = [T_1, T_2] = [T_2, T_3] = \ldots = [T_{n-1}, T_n] = T$. As a consequence, its ownership cost C_p is given by:

$$C_p = \frac{Q}{2} * P_u * i \qquad (6.5)$$

And the materials cost is given by:

$$C_m = K * P_u \qquad (6.6)$$

As the three right terms are well known from the formulas above, Equations (6.3), (6.5), and (6.6), then, from Equation (6.1), the stock's total annual cost can be calculated:

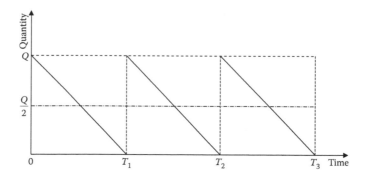

FIGURE 6.2
Stock variation in time.

$$CT = C_{at} + C_m + C_p \tag{6.1}$$

$$\Leftrightarrow$$

$$CT = C_a * \frac{K}{Q} + K * P_u + \frac{Q}{2} * P_u * i \tag{6.7}$$

From Equation (6.7), it may be seen that it is possible to find the optimum time between orders with the optimum economic amount of stock orders Q_e, which happens when CT is minimal. This calculation is made by setting the derivative of CT with respect to Q to zero.

$$\frac{dCT}{dQ} = 0 \Leftrightarrow \frac{dCT}{dQ} = -\frac{K * C_a}{Q^2} + 0 + \frac{1}{2} * P_u * i = 0 \Leftrightarrow \frac{K * C_a}{Q^2} = \frac{1}{2} * P_u * i \Leftrightarrow Q^2 = \frac{2 * K * C_a}{P_u * i} \tag{6.8}$$

Finally, this yields:

$$Q = \sqrt{\frac{2 * K * C_a}{P_u * i}} \tag{6.9}$$

The preceding formula is referred to as the Wilson formula and permits determination of the economic order quantity of a material in the case of periodic orders and assuming a linear variation of the stock in time.

$$Q_e = \sqrt{\frac{2 * K * C_a}{P_u * i}} \tag{6.10}$$

Equation (6.10) is called the Wilson formula. The original model was developed by F. W. Harris in 1913, but it was R. H. Wilson who began its extensive application after the initial model was developed. This is the reason for the assignment of his name to the final formula.

The economic order quantity Q_e will be ordered N times a year with an optimum interval T_e between orders:

$$T = \frac{1}{N} = \frac{1}{\frac{K}{Q}} = \frac{Q}{K} \Leftrightarrow T_e = \frac{Q_e}{K} = \sqrt{\frac{2 * C_a}{P_u * i * K}} \tag{6.11}$$

The following example clarifies the use of the method.

Knowing that the annual consumption of a certain spare part is 55 units ($K = 55$), the administrative acquisition cost of ordering of 100 CU ($C_a = 100$ CU) (CU—Cost Unit), the unit price of 20 CU ($P_u = 20$ CU), and the possession rate of the average value of the annual stock, 15% ($i = 15\%$), calculate the quantity of items to be ordered in each restock (Q) and the time interval that must occur between two successive orders (T).

From the immediate application of Wilson's formula [Equation (6.10)]:

$$Q_e = \sqrt{\frac{2*55*100}{20*0,15}} \cong 60,55 \cong 60\,\text{Units}$$

And the economic time interval T between orders [Equation (6.11)] is the following:

$$T_e = \frac{Q_e}{K} = \frac{60}{55} \cong 1,09 = 1\,year + 0,09*12 = 13\,months$$

On this relevant and difficult subject, particularly to discuss the remaining three situations, the following references may help a lot: Orsburn (1991), Gopalakrishnan and Banerji (2004), and Slater (2010).

6.3.1 Other Approaches

According to Table 6.3, there are more three situations to manage the two main stock variables, that is, time and quantity, such as the following:

- Order point (quantity periodic, time variable)
- Procurement plan (quantity variable, time periodic)
- Security parts (quantity variable, time variable)

The order point method is based on the principle that a (predetermined fixed-size) order is placed whenever the stock falls to a preset level, called the order point. As demand is higher or lower, this point is reached more or less rapidly. As a consequence, this is a method where the quantity to be ordered is fixed and the time between orders variable.

This method implies continuous knowledge of the existence of and, as a consequence, a close control on all stock movements.

The occurrence of ruptures depends on the demand behavior during the replacement time, such as:

- A high demand during the replacement time, resulting from two factors:
  - Exceptional consumption
  - A longer-than-expected replacement time
- The replacement time, which plays a very important role in the performance of stock management and in the investment value of the stock needed to provide an adequate answer to demand

The procurement plan, also called the cyclical review method, is characterized by the fact that orders are made at fixed intervals and the quantity to be ordered is variable, calculated in order to raise the existing stock plus the quantity ordered to fill the stock to a maximum level.

In this method, there is no point in the cycle in which the value of the physical stock number is known *a priori*. However, at the points when the inventory is reviewed and, as a consequence, an order is made, the existing stock plus the ordered stock reaches a value that is identical in all cycles. This stock value will have to cover all the needs until the moment the order is placed at the next cycle time.

The security parts value corresponds to the portion of the stock reserved to respond to possible demand variations, depending directly on the standard deviation of the demand sample.

The models applicable to the security parts are aligned with other methods, namely order point and procurement plan. These are the main reasons the calculation of the security parts between orders will have a time interval and a variable quantity.

The security stock represents a number of parts in addition to the normal stock in order to avoid ruptures that, if they occur, may represent undesirable effects, be it in terms of costs, risks, production failures, or others. Security spare parts are critical to the equipment's availability, especially the A-class ones.

The cost of this type of stock is related to the desired security, and consequently a balance must be achieved between the cost of materials, the cost of storage, the cost of the stock rupture, and the assumed risk.

6.3.2 Pareto Analysis

The Pareto method, or ABC analysis, is a very simple management tool, but it has great effectiveness in stock classification. This method classifies the stocks into three large groups, A, B, or C, according to the percentage of importance of each of these classes, such as the annual consumption, or costs, that each group represents.

Separation among the classes is done according to the following methodology:

- Class A—Corresponds to the most important group of materials—although they are represented by a small number of materials, 15%–20% of the total, they correspond to 75%–80% of the cost, or other relevant criteria, of the total annual consumption.
- Class B—This is an intermediate group, where 20%–25% of the total articles represent 10%–15% of the cost, or other relevant criteria, of the total annual consumption.
- Class C—This group of articles has the lowest value of annual consumption, although it represents a high number of references, about 60%–65% of the total number of materials corresponding to 5%–10% of the total annual consumption.

The management of each class may be carried out as follows:

- Class A—The materials should be checked frequently in order to maintain low stock levels and simultaneously prevent breaks.
- Class B—The materials should be controlled in a more automated way, that is, with less detail.
- Class C—The materials must have very simple and fully automated decision rules—the safety stock levels of this class can be raised to minimize the drawbacks of any breaks.

The ABC methodology algorithm is the following:

a. To order the materials in descending order of quantitative importance

b. To calculate the accumulated value

c. To calculate the percentage of each material

d. To calculate the accumulated percentage

e. To calculate the cumulative percentage of number of references

f. To rank the materials (A, B, or C)

The case study in this chapter refers to a spare part analysis in the maintenance department of a European automotive company.

6.4 Tools

Tool management can be discussed in two ways:

a. The control of tools themselves

b. Special tool control, namely the equipment for tests and measurements

Chapter 7, in Section 7.3.4 ("Tools"), discusses the form of a CMMS, as shown in Figure 7.9 ("Tools form"), as a way to control tool resources.

According to the first item, the control of tools themselves is important because the investment in the usual tools is usually very high and a lot of them usually disappear over time. Thus, it is important to know to whom each individual tool belongs. This approach implies an additional responsibility for each tool that is indexed to the owner, which usually implies an additional responsibility in its use and treatment.

Additionally, it is necessary to manage the special tools carefully, because there are usually only one or two units of each one and, as a consequence, working orders must be launched taking these restrictions into account. These special tools are very expensive and, as a consequence, must be managed carefully, namely for planned interventions, in order not to overschedule their use for two or more working orders simultaneously.

Special tools, for example, vibration tools, thermography, ultrasound, and/or other equipment with this specificity, have the restrictions mentioned in the preceding paragraph and, in some situations, a cost per use may be allocated in each working order. It is because of this that working control, management, and closing control are so important, as can be seen in Section 5.3 and in Figure 5.4.

6.5 Case Study

The case study presented here is taken from a European automotive company. The data, both in this section and throughout this book, are real. However, for confidentiality reasons, the real names of both the company and materials are omitted.

As can be seen from the grey cells in Table 6.4, this company has a well-defined ABC distribution in its maintenance materials. This distribution, even appearing as theoretical, in fact represents many decades of experience in maintenance management.

Figure 6.3 shows the graphical ABC distribution.

TABLE 6.4

Stock Spare Parts in an Automotive Company

Materials/Code	No. of Items	%	% Accumul.	Materials/Code	No. of Items	%	% Accumul.
(Part I)							
D87	4152	15.76	15.76	X48	163	0.62	85.03
X00	1518	5.76	21.53	X15	158	0.60	85.63
X75	1480	5.62	27.14	X81	157	0.60	86.23
X37	1392	5.28	32.43	X91	151	0.57	86.80
P95	1065	4.04	36.47	X06	145	0.55	87.35
X34	1016	3.86	40.33	P68	141	0.54	87.89
X25	939	3.56	43.89	P93	140	0.53	88.42
X30	935	3.55	47.44	X42	125	0.47	88.89
X04	927	3.52	50.96	X03	122	0.46	89.35
P88	719	2.73	53.69	T85	119	0.45	89.81
X65	705	2.68	56.37	T84	118	0.45	90.25
X41	493	1.87	58.24	X27	116	0.44	90.70
X60	458	1.74	59.98	X61	114	0.43	91.13
X23	442	1.68	61.66	T80	111	0.42	91.55
T10	407	1.55	63.20	P76	103	0.39	91.94
X36	394	1.50	64.70	P08	97	0.37	92.31
X11	348	1.32	66.02	X54	95	0.36	92.67
P67	332	1.26	67.28	X14	93	0.35	93.02
X12	318	1.21	68.49	T94	92	0.35	93.37
X51	314	1.19	69.68	X17	91	0.35	93.72
X95	297	1.13	70.81	P25	78	0.30	94.01
X72	296	1.12	71.93	P64	73	0.28	94.29
X01	289	1.10	73.03	X43	70	0.27	94.56
X79	274	1.04	74.07	T78	68	0.26	94.81
X74	264	1.00	75.07	X77	67	0.25	95.07
X85	261	0.99	76.06	X89	66	0.25	95.32
X62	261	0.99	77.05	X84	63	0.24	95.56
X64	250	0.95	78.00	X83	58	0.22	95.78
X68	250	0.95	78.95	P31	57	0.22	95.99
X70	244	0.93	79.88	P13	52	0.20	96.19
X21	235	0.89	80.77	P91	52	0.20	96.39
X10	225	0.85	81.62	P58	51	0.19	96.58
P02	196	0.74	82.37	T86	46	0.17	96.76
X19	191	0.73	83.09	P66	41	0.16	96.91
X32	182	0.69	83.78	T72	38	0.14	97.06
X57	166	0.63	84.41	T22	38	0.14	97.20
(Part II)							
P19	36	0.14	97.34	X93	6	0.02	99.67
P55	36	0.14	97.48	P15	6	0.02	99.70
X97	33	0.13	97.60	X18	6	0.02	99.72
T53	33	0.13	97.73	T52	6	0.02	99.74
P78	31	0.12	97.84	T20	5	0.02	99.76
T46	30	0.11	97.96	T31	5	0.02	99.78
X45	29	0.11	98.07	P34	5	0.02	99.80

(Continued)

TABLE 6.4 (*Continued*)

Stock Spare Parts in an Automotive Company

Materials/ Code	No. of Items	%	% Accumul.	Materials/ Code	No. of Items	%	% Accumul.
T00	28	0.11	98.17	X47	4	0.02	99.81
T19	24	0.09	98.27	X67	4	0.02	99.83
P48	23	0.09	98.35	P50	4	0.02	99.84
T49	22	0.08	98.44	T70	4	0.02	99.86
P97	20	0.08	98.51	T73	3	0.01	99.87
X55	20	0.08	98.59	T51	3	0.01	99.88
P44	19	0.07	98.66	P06	3	0.01	99.89
P39	19	0.07	98.73	X52	3	0.01	99.91
P62	18	0.07	98.80	P75	2	0.01	99.91
T75	17	0.06	98.86	P61	2	0.01	99.92
P32	17	0.06	98.93	X40	2	0.01	99.93
X98	16	0.06	98.99	T71	2	0.01	99.94
T01	14	0.05	99.04	P71	2	0.01	99.94
T15	14	0.05	99.10	P09	1	0.00	99.95
X38	14	0.05	99.15	T66	1	0.00	99.95
T11	14	0.05	99.20	P42	1	0.00	99.95
X71	13	0.05	99.25	P36	1	0.00	99.96
T23	12	0.05	99.30	T60	1	0.00	99.96
T95	11	0.04	99.34	D83	1	0.00	99.97
X87	10	0.04	99.38	X90	1	0.00	99.97
P00	10	0.04	99.42	T88	1	0.00	99.97
P04	10	0.04	99.45	X76	1	0.00	99.98
T68	9	0.03	99.49	T89	1	0.00	99.98
P73	8	0.03	99.52	X33	1	0.00	99.98
T82	7	0.03	99.54	X39	1	0.00	99.99
P40	7	0.03	99.57	P98	1	0.00	99.99
T44	7	0.03	99.60	T26	1	0.00	100.00
X49	7	0.03	99.62	P57	1	0.00	100.00
T16	7	0.03	99.65	**Grand Total**	26341	100.00	100.00

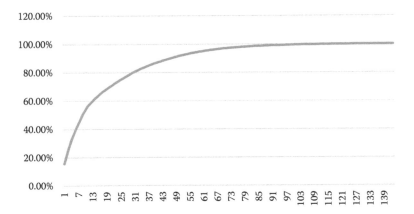

FIGURE 6.3

ABC analysis referred to in data of Tables 6.4 I and II.

7

Integrated Systems for Maintenance Management

7.1 Background

Maintenance activity in all its scope is extremely complex, due to the diversity inherent in the various subjects that constitute it and also to the combination of them.

With the advent of information systems, many aspects of maintenance activities have evolved, notably due to the increase in research and development (R&D) carried out. As a result of this, tools that are available to users were created, such as integrated systems for maintenance management, usually called computerized maintenance management systems and, nowadays, enterprise asset management systems. The last ones include the first and also the entire life cycle of physical assets, that is, from acquisition until withdrawal.

EAMs/CMMSs have great diversity and even have many similar aspects, such as those involving the management of the main areas of maintenance, which includes much common data that support and feed those systems.

According to these characteristics, it is important to know the structure of a maintenance information system, as well as the most relevant sets of data for maintenance management.

This is the subject addressed in this chapter, in which the Integrated Modular System of Terology, created by the author of this book, will be used as a reference. The objective is to use a real system as a case study that can be extrapolated for similar systems (Farinha, 1994, 1997; Farinha et al., 2004, 2010).

7.2 Software and Hardware Options

Nowadays, there are a lot of options for implementing a CMMS. However, the tendency is to use the cloud to host the program and data and a personal computer, tablet, and/or smartphone to access the system.

The traditional systems, called server–client, implied that companies had a server, many times a small data center, as the access to an internal network using personal computers. These systems usually contained a wire network that had restrictions to accessing the system. With the advent of wireless networks, the front ends can be of any type, that is, computers, tablets, and smartphones can be the access to the system through a browser or other means. These types of systems imply that companies invest a lot in hardware, namely in their own data centers. In these situations, the software is usually acquired with a contract for maintenance and upgrades.

As noted initially, the tendency is to use cloud services as a front end for any device, paying a license, many times monthly. Cloud computing is usually called SaaS. It helps eliminate large investments in hardware and software, as well as the need for major information technology resource involvement.

However, cloud computing also has some disadvantages, namely the following:

- Makes small businesses dependent on the reliability of the Internet connection. When it's offline, the company is offline.
- Cloud computing means Internet computing. This signifies that, if the company intends to protect its data, this solution may involve some fear. In fact, there is frequently news about access via piracy to what is considered the most secured data.
- Cloud computing applications may appear to be cheaper than a software solution installed and run in-house. However, the user can only choose the modules the supplier offers, and cannot adapt the software to his or her particular requisites in this case.
- The inflexibility of some cloud applications can be a serious disadvantage. It is necessary to be careful with applications and/or data formats that do not allow easy transference and/or conversion of data into other systems.
- Customer support may be another weakness of cloud computing. It is necessary to guarantee an adequate response to customer support issues. The technical support may include e-mail, phone, live chat, knowledge bases, and user forums.

7.3 Structure of Information Systems for Maintenance

To discuss the structure of an information system for maintenance, the author's information system, called the Integrated Modular System of Terology (SMIT), is used as a reference. It has the following characteristics:

- It is an integrated modular system, which means that it is an information system developed in a modular way, integrating several modules required to manage asset maintenance and with the capacity to integrate new modules.
- The forms can work in any language, adapting themselves to the users' choice.
- Terology is the concept behind it, as described in the opening chapters, which means the managing of all assets' life cycles.

SMIT was initially implemented with PHP and PostgreSQL (Farinha et al., 2008). Nowadays, it also works in the cloud, as SaaS, and with any type of device and browser. The SMIT logo is shown in Figure 7.1 and the main modules in Figure 7.2.

The main modules of SMIT are the following (Figure 7.2):

- Physical Assets (PAs)/Maintenance Objects
- Customers
- Suppliers
- Technicians

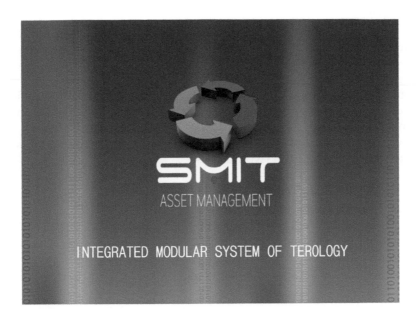

FIGURE 7.1
SMIT logo.

- Tools
- Spare Parts
- Work Orders
- Intervention Requests
- Fault Diagnosis
- Maintenance Plans
- Maintenance Contracts
- Acquisition and Withdrawal

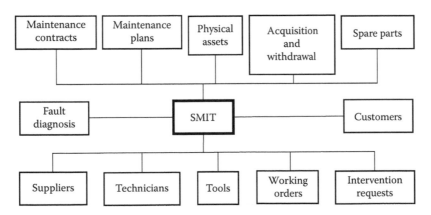

FIGURE 7.2
Main modules of SMIT.

7.3.1 Maintenance Objects/Physical Assets

Integrated maintenance management systems may apparently assume several forms and approaches, depending on the profile of the project team and the business sector they came from, among others.

The system presented here, SMIT, has the modules described in the previous section, with its architecture supported by the following main pillars: physical assets, work orders, and maintenance planning.

The Maintenance Objects module—Physical Assets—as well the other modules presented in this chapter will be analyzed in two ways:

- By the various data sets that constitute it
- By its interrelations with other modules

Figure 7.3 shows the form of Maintenance Objects and one of its sections (Physical Characteristics).

The various data sets included in the Maintenance Objects module are structured in the following records:

- Physical Characteristics
- Manuals
- Functional Characteristics

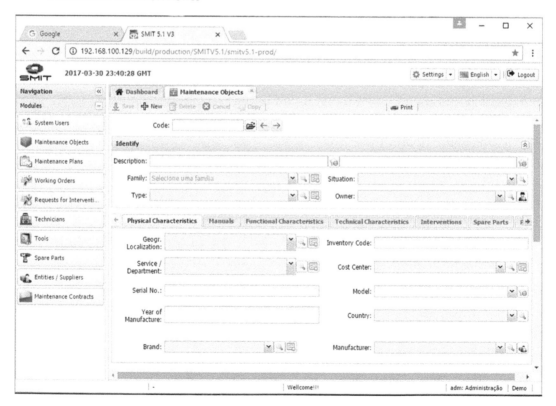

FIGURE 7.3
Form of MO module.

- Technical Characteristics
- Interventions
- Spare Parts
- Photographs
- Customers

These data groups aggregate information according to the theme of each section, as described in the preceding paragraph.

The current version of SMIT permits the user to implement a MO structure with the level of discrimination he or she wants, without limitations for the decomposition level, according to the real structure of the physical asset.

The main interrelations of the MO module and the other modules are illustrated in Figure 7.4.

The MO module has interrelations with almost all the other modules. This module is, as indicated above, one pillar of the system. The interrelationships are the following:

- Suppliers:
    - Of MO
    - Of spare parts
    - Of maintenance services
    - Of tools
- Spare Parts:
    - For planned interventions
    - For nonplanned interventions
- Technicians:
    - Who execute maintenance services recorded in WOs
- Working Orders:
    - Planned
    - Nonplanned

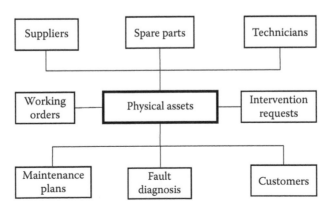

FIGURE 7.4
Interrelationships of MO module and other modules.

Asset Maintenance Engineering Methodologies

- Intervention Requests:
  - Requested by the users of MO
  - Requested by the technicians
- Maintenance Plans:
  - Systematic:
    - Periodic
    - Nonperiodic
  - Conditioning
- Fault Diagnosis:
  - From the data collected from manuals
  - From the data collected from the nonplanned WOs
  - From the knowledge collected from technicians
- Customers:
  - Who are assigned the costs of planned interventions
  - Idem, for unplanned interventions

7.3.2 Suppliers

The Supplier module (Figure 7.5) refers to providers of:

- MO/assets
- Spare parts

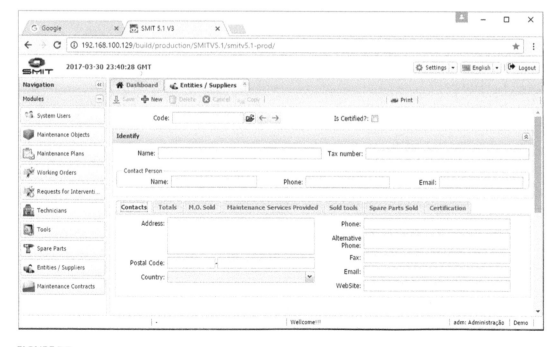

FIGURE 7.5
Suppliers form.

- Maintenance services
- Tools

This module has the following data groups:

- Total quantitative data
- Physical assets sold
- Maintenance services sold
- Tools sold
- Spare parts sold
- Certification

The Suppliers module's interrelations with other modules are the following (Figure 7.6):

- Maintenance Objects:
  - Sold by the supplier
  - Maintenance services sold
- Spare Parts:
  - Sold by the supplier
- Tools:
  - Sold by the supplier
- Work Orders:
  - Executed by a supplier

7.3.3 Technicians

The Technicians module (Figure 7.7) manages data related to maintenance technicians and is structured with the following data groups:

- Vacations/absences
- Tools
- Price/hour
- Maintenance services
- Costs

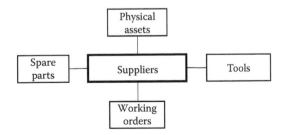

FIGURE 7.6
Interrelationships of Suppliers and other modules.

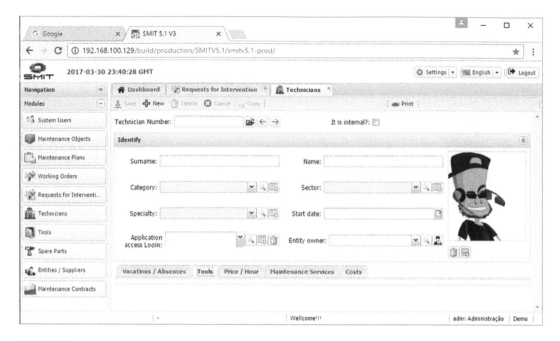

FIGURE 7.7
Technicians form.

The relations among the Technicians module and the other modules are illustrated in Figure 7.8.

The interrelations among the Technicians module and the other modules are the following:

- Work Orders:
  - Where the technicians participate
- Tools:
  - For which the technicians are responsible

7.3.4 Tools

The Tools module (Figure 7.9) manages tools data, which corresponds to the tools characterization, and the association with the technicians that is responsible for each tool.

The relations among the Tools module and the other modules are illustrated in Figure 7.10.

The interrelations among the Tools module and the other modules are the following:

- Technicians:
  - That have tools under their responsibility
- Work Orders:
  - Where they are used

FIGURE 7.8
The interrelations among technicians module and other modules.

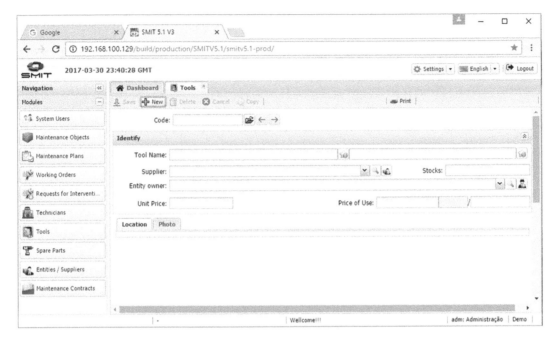

FIGURE 7.9
Tools form.

FIGURE 7.10
Interrelations among Tools module and other modules.

7.3.5 Spare Parts

The Spare Parts module (Figure 7.11), as the name suggests, refers to the spare parts of the MO, relevant to:

- Maintenance objects
- Maintenance plans
- Work orders, both planned and nonplanned

This module is composed of the following data groups:

- Stocks
- Order forecast
- Planned orders
- Alternative suppliers
- Manuals
- Photographs
- Booking/spare parts

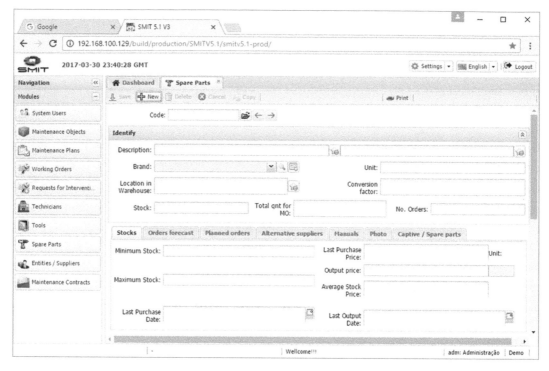

FIGURE 7.11
Spare Parts form.

The Spare Parts module's interrelations with the other modules are illustrated in Figure 7.12. The interrelations among the Spare Parts module and the other modules are the following:

- Maintenance Objects:
  - Spare parts indexed to each MO
- Maintenance Plans:
  - Spare parts programmed for each planned intervention
- Work Orders:
  - Planned:
    - Programmed spare parts

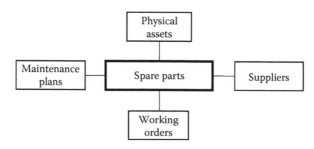

FIGURE 7.12
Interrelationships among Spare Parts module and other modules.

- Not-planned:
  - Used spare parts
- Suppliers:
  - Of spare parts
  - Alternative suppliers

7.3.6 Work Orders

The Work Orders module (Figure 7.13) manages the data for one of the main modules of the information system, as initially referred to. The proper monitoring and control of the WO module depends on the feedback of the system in crucial aspects such as the following:

- Maintenance planning
- Cost control
- KPI calculation
- Fault diagnosis
- History

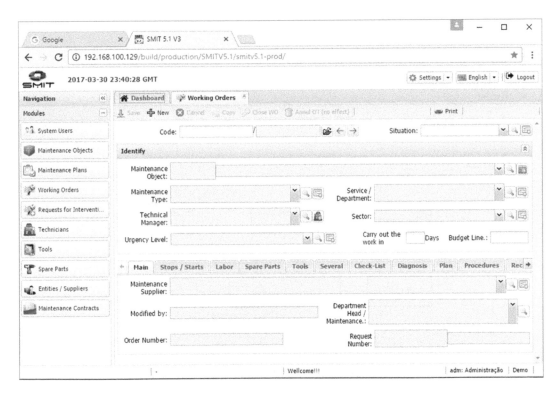

FIGURE 7.13
WO form.

This module is composed of the following data groups:

- Main data
- Stops/starts
- Labor
- Spare parts
- Tools
- Procedure checklists
- Fault diagnosis
- Planning
- Procedures
- Costs

The interrelations of the Work Orders module and the other modules are illustrated in Figure 7.14.

The interrelations among the Work Orders module and the other modules are the following:

- Spare Parts:
  - Applied in maintenance interventions
- Maintenance Objects:
  - The ones to which maintenance intervention refers
- Intervention Requests:
  - That require the opening of a WO
- Technicians:
  - Who worked on each WO
- Tools:
  - Used in the intervention of the WO
- Fault Diagnosis:
  - In the case of unplanned WOs

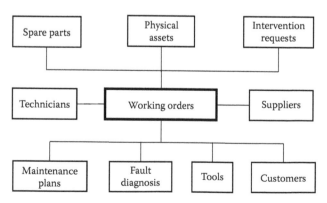

FIGURE 7.14
Interrelations among WO module and other modules.

- Suppliers:
  - For outsourcing interventions
  - For spare parts
- Maintenance Plans:
  - That originated the WO
- Customers:
  - To whom the costs of the WO are allocated

7.3.7 Intervention Requests Module

The Intervention Requests module (Figure 7.15) allows a request for assistance to the maintenance department by the users of the MO or the technicians. From this module, the working orders (WOs) generated are not necessarily planned and, in a breakdown situation, it gives the first information about the occurrence. Some crucial aspects of this module are the following:

- Unplanned WO
- Fault diagnosis

This module is composed of the following data groups:

- Registration time
- Occurrence/comment maintenance department
- Add/change information to request

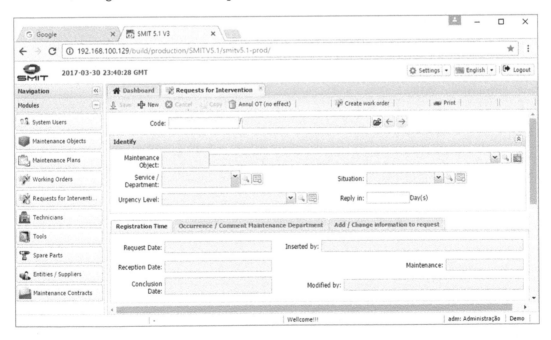

FIGURE 7.15
Intervention Requests form.

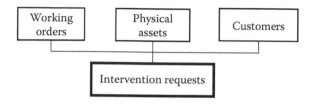

FIGURE 7.16
Interrelations of Intervention Requests module and other modules.

The interrelations among the Intervention Requests module and the other modules are illustrated in Figure 7.16.

The interrelations among the Intervention Requests module and the other modules are the following:

- Maintenance Objects:
  - Referred to in the intervention request
  - Which data intervention is recorded
- Work Orders:
  - Not planned
- Customers:
  - The service/department in which the MO was allocated at failure time
  - For which the costs will be shared

7.3.8 Fault Diagnosis Module

The Fault Diagnosis module, as its name implies, allows diagnosis of the faults in the MO. The data in this module originated in:

- Unplanned WOs
- Service manuals
- Technicians knowledge

The Fault Diagnosis module can be accessed through a specific module or in the WO module in a specific section, as shown in Figure 7.17.
This module has the following data groups:

- Occurrence
- Cause
- Procedure
- Key words

The interrelations of the Fault Diagnosis module and the other modules are illustrated in Figure 7.18.

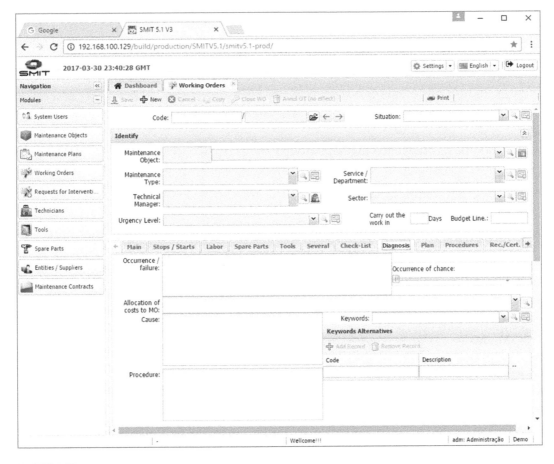

FIGURE 7.17
Fault Diagnosis form into WO.

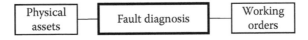

FIGURE 7.18
Interrelations of Fault Diagnosis module and other modules.

The interrelations among the Fault Diagnosis module and the other modules are the following:

- Maintenance Objects:
  - Referred-to diagnostic data
- Working Orders:
  - Where the diagnostic data came from
  - Where existing diagnostic data can be queried

7.3.9 Planning

The Maintenance Plans module (Figure 7.19) allows intervention planning, both periodic and aperiodic. The data sources in this module are from:

- Service manuals of the manufacturers
- Technicians' knowledge
- WO history

This module has the following data groups:

- Planning
- Information about revisions
- Human resources
- Spare parts
- Tools
- Procedure checklists
- Actions
- Action support
- Manuals
- Expected costs

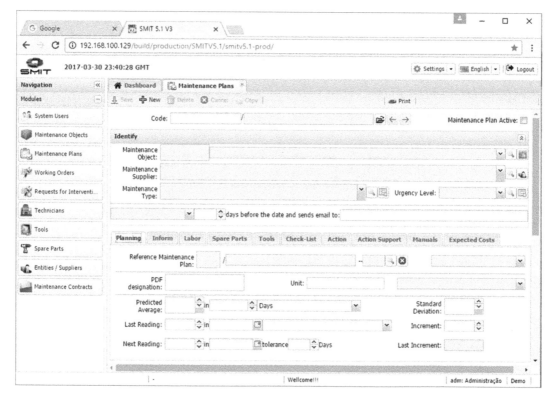

FIGURE 7.19
Maintenance Plans form.

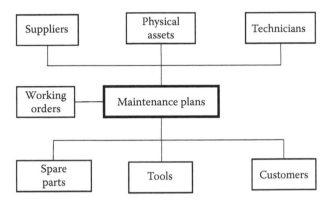

FIGURE 7.20
Interrelationships of Maintenance Plans module and other modules.

This module works in conjunction with a Gantt chart tool, which permits adaptation of maintenance plans to the specific constraints of each moment.

The interrelationships of the Maintenance Plans module and the other modules are illustrated in Figure 7.20.

The interrelationships of the Maintenance Plans module and the other modules are the following:

- Maintenance Objects:
  - For which the planning is done
  - The indexation of one plan to others because of calendar restrictions
- Suppliers:
  - Who execute the planned WO
- Technicians:
  - Who execute the planned WO
- Spare Parts:
  - Necessary for the planned interventions
- Tools:
  - Necessary for the planned interventions
- Customers:
  - Of the MO
- Work Orders:
  - Issued to the corresponding maintenance plans

7.3.10 Customers Module

The Customers module allows management of the data of customers that use the MO. The data source of this module is:

- Customers who purchase products or services that use the MO

This module is composed of the following data groups:

- Total cost of MO
- Maintenance services
- Certification
- Associated MOs

The interrelations among the Customers module (Figure 7.21) and the other modules are the following:

- Maintenance Objects:
  - Associated with customers
- Intervention Requests:
  - For MO with customers associated
- Maintenance Plans:
  - With customers associated
- Work Orders:
  - Of MOs associated to customers

7.3.11 Condition Monitoring

Condition monitoring is, more and more, the most adequate approach to extend the intervals between maintenance interventions and, as a consequence, to maximize equipment availability.

There are many variables that can be managed to accompany health equipment, like the following:

- Vibrations
- Oils
- Effluents
  - Exhaust car systems
  - Industrial chimneys
  - Others
- Temperatures
- Others

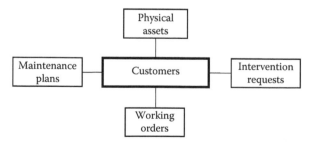

FIGURE 7.21
Interrelations of Customers module and other modules.

From the accompanying of the condition variables and through prediction tools, the next intervention can be planned before the correspondent variable reaches its limit instead of making interventions based on a periodic schedule that is usually on a shorter time.

Condition monitoring is described in detail in Chapter 12. The reason it is referred to in this section is that data reading from sensors ought to be managed in a CMMS system like the one described in this chapter.

Whether the data are read manually or through sensors, they finally ought to be used to manage the next interventions, with the objective of the enlargement of the time between interventions as much as possible and also to diminish the repair time. The final objective is to maximize the equipment's availability.

7.3.11.1 Sensor Reading

To implement a condition monitoring policy, it is necessary to select adequate variables and whether the reading will be made online or offline. There are several questions that must be analyzed in order to decide what type of solution best suits each specific situation.

For online condition monitoring, it is necessary to evaluate the real conditions for implementing it, namely the following:

- Investment cost for implementing the sensors
- Communication lines, wire and/or wireless
- Electrical power sources
- Time interval between readings
- Hardware resources
- Software resources

For offline condition monitoring, it is necessary to evaluate the real conditions for implementing it, namely the following:

- Investment cost for implementing the system
- Equipment for measurement and registry
- Interval between measurements
- Reading logistics
- Human resources necessary
- Software tools for analyzing data

This subject is developed in detail in Chapter 12.

7.3.11.2 Technological Options

The technological options for condition monitoring are diverse, as can be seen in detail in Chapter 12. The reason this subject is referred to here is, as mentioned before, the CMMS must be connected to condition monitoring systems. However, there are no standard systems, protocols, and so on that transparently connect them.

This subject is extremely sensitive, which is why it is referred to here and later, as mentioned. In fact, for example, the choosing of a thermographic camera or vibration equipment includes the decision of what type of communication they have (Bluetooth, USB, etc.) and also the type and format of data they transfer and for which support (Excel, SQL database, etc.).

If this type of question isn't correctly answered, we can have heterogeneous tools for test and measurement, which implies much more difficult and costly development of software tools to treat and manage the data sent by those tools.

The preceding questions are applicable to situations with online reading. In this case, the situations are much more diverse, because each manufacturer uses its own systems and protocols. Additionally, many times, there are also hardware and software developed by the owner of the equipment to extend the options for online sensor reading. Some questions to be considered are the following:

- What type of sensors are necessary?
- What type of hardware and software are necessary to read and manage the data?
- What type of transmission is necessary—wire or wireless?
- What type of format of data and what type of software to receive the data are needed?

These types of questions are strategic because the data must be managed, both for condition monitoring and also to predict the next interventions. The treatment of data can be made autonomously or automatically by the CMMS.

The transversal problem to solve is the following:

- We must have data tools that all work in a similar way, both in transmission and data format, in order to simplify the data storage and following treatment as much as possible. The final objective is always the same, to predict the next intervention, and it is here that the optimization efforts must be placed.

7.4 A Computerized Maintenance Management System/Enterprise Asset Management Example

This section presents an example of implementation of a CMMS/EAM in a little hospital of 52 beds in a European country. The steps taken will be described in order to implement the system and put it in production, and also cover some reorganization aspects.

The main chronology is the following:

- Making dossiers of general equipment
  - Codification
  - Hierarchical structure
  - Characteristics
  - Complementary data
- Making dossiers of medical equipment
  - Codification
  - Hierarchical structure
  - Characteristics
  - Complementary data

- Designing the maintenance plans
- Issuing and controlling the nonplanned working orders
- Launching and controlling the planned work orders
- Managing the outsourcing contracts
- Launching and controlling the calibration process
- Analyzing the LCC

The first step of implementing the CMMS, making dossiers of general equipment, begins with the decision concerning which physical assets to start with. This was the heating, ventilation, and air conditioning (HVAC) system.

The next step was to define what type of codification to use. The European article number (EAN) with 13 digits with a bar code system, EAN 13, was adopted. The reasons were the following: the million possibilities (physical assets) that can be used inside the organization, the validation by check-digit, and the multiplicity of commercial support that exists (tags in paper, plastic, and others, and also laser and mechanical recording).

Following this decision, the hierarchical structure and the colocation of one tag in each module of the hierarchy of the HVAC system were made. This subdivision was decomposed into almost 500 modules, with the HVAC itself as the father equipment.

During this phase, photos of each module were taken, as well as the sign plate and its characteristics.

Finally, complementary data were collected, like manuals (service and operation, drawings, maintenance plans, and so on). Many elements were on-paper support and had to be digitalized.

In the next phase, making dossiers of medical equipment, the procedure was similar to the previous, but for equipment instead of general equipment.

Figure 7.22 shows a form with a module of equipment (a compressed air unit).

The next phase, designing the maintenance plan, is probably one of the most difficult and time intensive because of the enormous amount and diversity of equipment, both general and medical. In this phase, the procedures, spare parts, human resources, and times between interventions must be defined. The following procedures are with respect to medical equipment for ophthalmology (ophthalmic surgical microscope):

- Yearly
  - Visual and functional control of all equipment.
  - Visual control of all cables.
  - Check electrical safety.
  - Check the lifting systems, especially if the bearings are tight.
  - Check electrical safety according to IEC 62353.
  - Check chassis current and leakage.
  - Check control elements as handles.
  - Dust fan/filter: Check ventilation and suction power.
  - Check the optical image and field of view illumination.
  - See if it is possible to adjust the focus, zoom, and diaphragm of the field of view manually and freely.

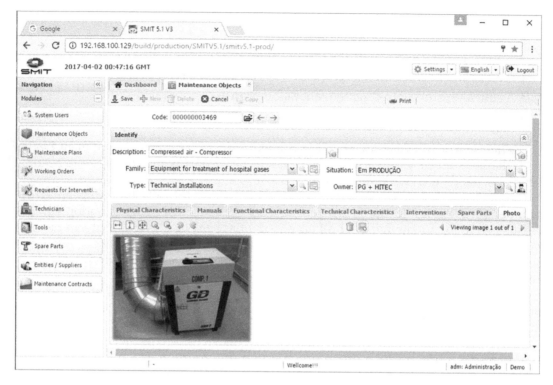

FIGURE 7.22
Form with a section of an equipment dossier.

- Biannually
    - Test optical conductor.
    - Test the lighting spectrum.
    - Test the light source filters.
- Every 4 years
    - Replacement of casters.
    - Replace the optical conductor.

The next phase, issuing and controlling nonplanned working orders, can start before the preceding phase, because faults may occur when the phase of maintenance plan design occurs. However, if possible, the preceding phases ought to be fulfilled before working order management starts.

The phase launching and controlling the planned work orders can be started after the partial or, ideally, the total plan of physical assets is inserted into the CMMS. These planned WOs are launched periodically, for example, weekly, monthly, or other, according to the number of assets under management. Figure 7.23 shows a planned working order. Figure 7.24 shows a nonplanned WO.

The phase of managing the outsourcing contracts allows dealing with the maintenance suppliers. In the example organization for the implementation of the CMMS under

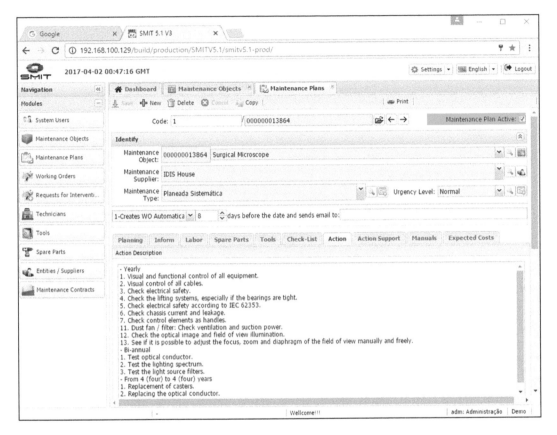

FIGURE 7.23
Form with a section of a maintenance plan.

discussion, the maintenance interventions are mostly outsourced. Figure 7.25 shows a section of a form of an outsourcing contract.

The phase launching and controlling the calibration process is pertinent, in this case of a hospital, only in specific situations. Thus, the example under discussion refers to this type of organization.

The management of calibration procedures can be done through a specific module or by the work orders module. In this case, it was done through this last solution in order to implement the CMMS as soon as possible.

The last phase, analyzing the LCC, allows following the LCC and evaluating, year after year, which is the most rational time to withdraw each piece of equipment. This analysis was done in detail in Chapter 3.

The total CMMS was implemented in 1 year, but it continues improving in order to consolidate its functioning and, according to KPI results, permit continuous improvement and, as is the objective of the hospital, implement and obtain an ISO 55001 certification.

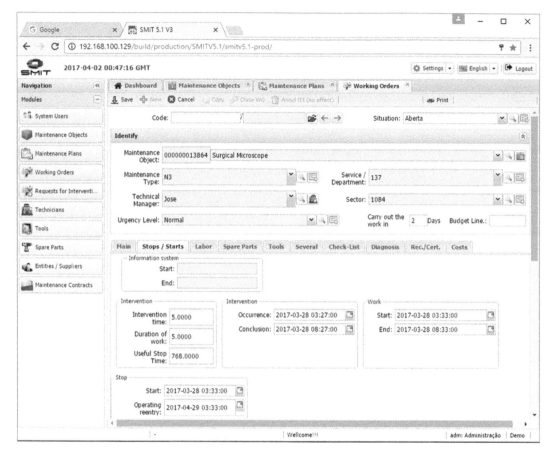

FIGURE 7.24
Form with a section of a nonplanned working order.

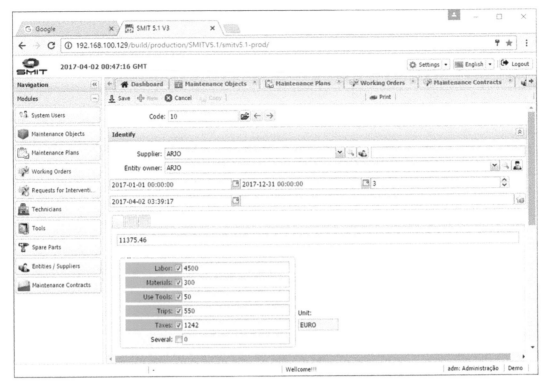

FIGURE 7.25
Form with a section of an outsourcing contract.

8

Expert Systems for Fault Diagnosis

8.1 Background

Expert systems for fault diagnosis are important tools to help the maintenance staff solve faults, especially the most difficult ones. This type of system has the advantages of helping to solve equipment faults with quality, within a rational time frame, usually less than when the technician tries to solve the problem without any aid.

Obviously, an expert, who knows the equipment and type of fault in detail, may solve the problem in a similar time, but many times, this particular situation doesn't work. However, if a company has an expert system for fault diagnosis, it can help if the human expert is not in the company or there is no technician who can solve the problem quickly and with quality.

Additionally, over time, the expert system will be enriched with more and more data from new faults, which significantly increases its performance.

This chapter describes fault diagnosis systems, supported by expert systems, with particular emphasis on the translation of uncertainty often associated with the observations and impact of such systems in this domain of application.

One approach emphasized in this chapter is supported in an inference process called case-based reasoning and fuzzy logic, which allow a general approach to fault diagnosis independently of specific equipment (Farinha et al., 2004; Marques, 2005; Pincho et al., 2006; Marques et al., 2009).

8.2 Profile of an Expert System for Fault Diagnosis

An expert system, commonly known as ES, is one of the products of the area of artificial intelligence (AI), whose objective is to solve problems in the most diverse domains, in which maintenance is a privileged field and is intended to reach results similar to those obtained by an expert.

The first ES was developed at Stanford University in the mid-1960s and was called DENDRAL. Its development led to the conclusion that an ES must operate in a specific domain of application and possess a substantial amount of knowledge about that domain. The term knowledge-based expert system (KBES), often used instead of ES, demonstrates this fact (Marques et al., 2001).

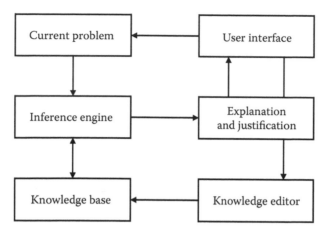

FIGURE 8.1
Basic structure of an expert system.

The areas of application of expert systems are diverse, such as the following:

- MYCIN—A classic of expert systems developed at Stanford University in the 1970s at the medical school to diagnose hospital infections. About 2,000,000 people/year in the United States are affected by such problems when they remain hospitalized. Of these, about 50,000 die.
- XCON—A system where the Digital Equipment Corporation (DEC) designs customized minicomputer configurations and manages order processing in the manufacturing and distribution sectors.
- CATS—A system that operates at General Electric (GE) to diagnose failures in diesel-electric locomotives, implemented in the 1980s to minimize GE's dependence on the services of its former chief engineer.

The basic structure of an ES can be represented by the diagram shown in Figure 8.1.

The interface with the user usually assumes a graphical form, and the interpretation and generation module use a natural language, which is another subject of study in artificial intelligence. The use of natural language is not mandatory but can be rewarding.

Another type of inference process used by ESs is called case-based reasoning (CBR), in which the previous cases are added to the knowledge base to support inferences about new cases.

The process of inferring conclusions and (possible) evaluation of their performance through the application of rules or determination of similarity between cases constitutes the motor of inference.

Adding cases, changing various parameters and definitions, and inserting them in the knowledge base are some of the tasks of the knowledge editor that uses the services of the user interface module.

The current problem can be described in several ways, but in the CBR process, it usually has an object-based structure that describes the current case. The diagnosis application can be simply constituted by the set of observations that describe the occurrence, which, after being related to the respective confirmed diagnoses, constitute the greatest part of the contents of the knowledge base.

The implementation of an ES can be based on several solutions, such as:

a. Rule-based ES-Shell—This name can be applied to a series of products, from simple languages specially designed for ES development, such as CLIPS, to a complete ES, but distributed with an empty knowledge base that will be filled according to the specific application domain where it will be applied, such as EXSYS, xPertRule, and EMYCIN.

b. Case-based ES-Shell—These are products of the same type as the previous, but for CBR systems, for example, ArtEnterprise, EasyReasoner, and Recall.

c. AI area languages (Prolog, LISP) and extension tools such as FLEX and FLINT.

d. Object-oriented/structured languages, such as Visual Basic and C++.

Solutions a) and b) have the objective of implementing rule-based and case-based systems, respectively, and there may be some overlap. Most of these correspond to commercial products with technical support, which permits fast implementation in the development of an ES. However, there are also shells commercially available, but they are dedicated to research work, given that the shells themselves are the result of research, but they are very powerful and effective tools. Some systems, such as EXSYS and ArtEnterprise, for example, provide modules designed to enable their ESs to have access via the Internet. The C Language Integrated Production System (CLIPS) was developed by the National Aeronautics and Space Administration (NASA) and is freely available on the Internet, having originated commercial products such as ECLIPSE and CLIPS/R2. The fuzzy version of CLIPS, from the National Research Council (Canada), allows working with fuzzy logic.

Solution c) is based on predicate logic. Languages such as Prolog, specializing in artificial intelligence applications, allow all possible conclusions to be obtained from the set of facts and rules that are declared. However, they become difficult to apply because of the difficulty in implementing a research strategy or other type of restriction that prevents a combinational explosion of rules until a goal is reached.

Solution d) is suitable for the implementation of CBR systems, because the cases adapt well to a representation through objects or frames (a concept similar to an object) and the relations between them. Because these languages are structured, implementing a rule-based ES is also possible. In the case of generic languages more basic than ES-Shell, they allow greater flexibility in implementing solutions, but with more work for development.

8.3 Rule-Based Expert Systems

Rule-based expert systems (also known as *expert systems*) are the simplest use of artificial intelligence. A rule-based expert system uses rules as the knowledge representation coded into the expert system.

In a rule-based expert system, much of the knowledge is represented as rules, that is, as conditional sentences relating statements of facts to one another. An expert system requires a knowledge base and an inference procedure.

The definitions of rule-based expert systems depend almost entirely on expert systems, which may be systems that mimic the reasoning of a human expert in solving a

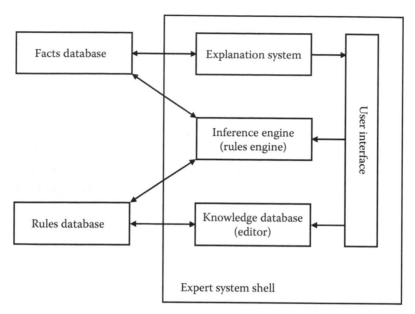

FIGURE 8.2
Generic rule-based expert system architecture.

knowledge-intensive problem. Instead of representing knowledge in a declarative, static way as a set of things that are true, rule-based expert systems represent knowledge in terms of a set of rules that indicate what to do or conclude according to each specific situation.

A rule-based expert system consists of a set of rules that can be repeatedly applied to a set of facts. The facts represent circumstances that describe a certain situation in the real world, as is the case with equipment faults. The rules represent heuristics that define a set of actions to be executed in a given situation.

However, the rule-based approach has several weaknesses, such as a poor capacity of generalization and handling novel situations. But it also offers efficiency and effectiveness for nondynamic systems operating within a fixed set of rules. In the modeling of the solving process, it can guide the user step by step in reaching the conclusion. Figure 8.2 shows the architecture of a generic rule-based expert system.

The decision trees correspond to the main technique to conduct the user to follow the most logical path to reach the most adequate conclusion. From the decision tree, the most relevant rules for each node can be written and the initial knowledge base can be constructed. Chapter 7, in Section 7.3.8 ("Fault Diagnosis Module") presents a form in Figure 7.17 ("Fault diagnosis form into WO") that is designed based on a decision tree.

According to Angeli (2010), rule-based systems do not require a process model. They require a multitude of rules to cover all possible faults and have difficulties with unexpected operations or new equipment. Among the several limitations of the many diagnostic expert systems, one must consider the inability to accurately represent time-varying and spatially varying phenomena, the inability of the program to learn from errors, and the difficulty of engineers acquiring reliable knowledge from experts, including technicians.

8.4 Case-Based Reasoning

Case-based reasoning is the process of solving new problems (cases) based on the solutions of similar past problems (cases). The operation cycle of a CBR system is well described by the Aamodt and Plaza diagram (Figure 8.3). Basically, it is composed of four phases:

1. Retrieve
   - Involves the search and selection of past cases, more or less similar to a new query case.
2. Revise
   - Makes a presentation of the (reused) solution and deals with its correctness or failure.
3. Reuse
   - May imply some kind of adaptation so that a past solution may be applied to the present case.
4. Retain
   - Records present cases classified as relevant for solving future ones.

This paradigm is extremely well adapted to the solution of problems like learning from experience, keeping experience available as needed, and quick knowledge transfer. However, CBR on its own is not enough, because, in the maintenance field, some know-how of the technical staff uses subjective experiences depending on visual inspection, noise, smell, and even approximate measurement of some attribute values. The translation of this kind of information is possible by means of fuzzy sets and fuzzy logic.

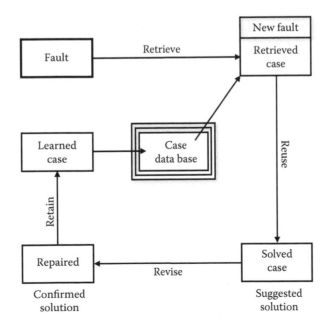

FIGURE 8.3
Aamodt and Plaza diagram adapted to fault diagnosis.

Based on the CBR process, expert systems for fault diagnosis can be developed, as is the case of SADEX, supervised by the author of this book and implemented by Marques (Marques et al., 2009).

In SADEX, the first element of an observation is the ID of the observed attribute. The system considers three attribute types: logical, measurable (e.g., temperature), and nonmeasurable subjective (e.g., smell). These three types give rise to two kinds of observations called absolute semantic observations and differential semantic observations. Some examples are the following:

1. The device does not work.
2. The temperature is equal to 10°C.
3. The temperature is low.
4. A burned smell is relevant.

These examples show that abnormality can be expressed in different ways. In fact, the maintenance teams, in their daily work, make effective use of these type of linguistic possibilities. Additionally, one of the important issues in the CBR paradigm is the global similarity computation between the query case and past cases that take place in the retrieve phase.

The software tools used to implement SADEX are standard and of low cost, allowing it to run on standard PCs.

8.5 Bayesian Models

Bayesian models are at the computational level, rather than at the algorithmic or process level, like traditional cognitive modeling paradigms. The algorithmic process may seem like a mechanistic procedure, but it may also require assumptions about human processing mechanisms. However, these are not needed when it is assumed that cognition is an approximately optimal answer to the uncertainty present in natural tasks and environments.

There are many reasons for choosing Bayesian methods—they have applications in several fields. Many professionals say that if a person wants to make consistent decisions in the face of uncertainty, the only way to do so is through the use of Bayesian methods.

One point that must be taken into account relates to logical problems with frequentist methods that do not arise in the Bayesian framework. Additionally, prior probabilities are intrinsically subjective, which is seen as a major drawback to Bayesian statistics. However, software tools allow Bayesian methods to manage large and complex statistical problems easily, where frequentist methods can only approximate or fail altogether. Bayesian modeling methods provide natural ways for people in many disciplines to structure their data and knowledge, and they yield direct and intuitive answers to real cases.

There are many varieties of Bayesian analysis. The fullest version of the Bayesian paradigm places statistical problems in the framework of the decision-making process. It involves the formulation of subjective prior probabilities to design pre-existing information; careful modeling of the data structure, checking and allowing for uncertainty in model assumptions; and formulating a set of possible decisions and a utility function to express how the value of each alternative decision is affected by the unknown model parameters.

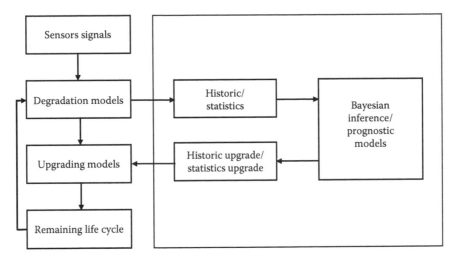

FIGURE 8.4
Generic Bayesian prognosis model.

However, each of the preceding steps can be omitted. Many cases of using Bayesian methods do not employ real prior information, either because it is weak or because they have high subjectivity, as is the case in the maintenance field—many pieces of equipment do have not historical fault information or, if it exists, it is not reliable.

Additionally, the decision theoretical framework is also frequently omitted, arguing that statistical inference should not really be formulated as a decision. In fact, there are varieties of Bayesian analysis and of Bayesian analysts. Figure 8.4 shows a generic Bayesian model based on a maintenance prognosis system fed by sensors placed in the equipment.

8.6 Data Mining

Data mining is the computer process of searching large stores of data to discover patterns and trends that go beyond simple analysis. Data mining uses mathematical algorithms to segment the data and, if necessary, predict future events. Data mining is also known as knowledge discovery in databases (KDD). The main properties of data mining are the following:

- Automatic discovery of patterns
- Prediction of likely outcomes
- Creation of manageable information
- Focus on large data sets and databases

Data mining corresponds to an approach that can answer questions that cannot be addressed through traditional query and reporting tools. The knowledge discovery process uses the following steps:

1. Identifying the problem
2. Data mining

3. Action

4. Evaluation and measurement

5. Deployment and integration into diagnosis processes

Knowledge discovery in databases process comprises some steps leading from raw data collection to some form of new knowledge, and its iterative process consists of the following steps (Figure 8.5):

- Data cleaning—Or data cleansing. It is a step in which the noise and irrelevant data are removed from the collection.
- Data integration—In this step, multiple data sources, even heterogeneous ones, may be combined in a common source.
- Data selection—In this step, the data relevant to the analysis are decided on and retrieved from the data collection.
- Data transformation—This is known as data consolidation. It is a step in which the selected data are transformed into forms appropriated for the mining procedure.
- Data mining—It is the crucial step in which some clever techniques are applied to extract potentially useful patterns.
- Pattern evaluation—In this step, only strictly interesting patterns representing knowledge are identified based on given measures.
- Knowledge representation—It is the final step in which the discovered knowledge is visually represented to the user; this critical step uses visualization techniques to help the user understand and interpret the data mining results.

In the maintenance field, data mining can play an important role in aiding fault diagnosis. For this, data classification is an important part of data mining process. Some common classification models include decision trees, neural networks, genetic algorithms, rough sets and statistical models, and so on. The decision tree algorithm is one of the most widely used in data mining algorithms.

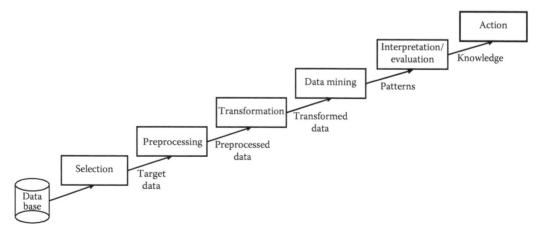

FIGURE 8.5
Iterative data mining process.

8.7 Performance Measures

As described in previous sections, there are several approaches to expert systems that pose a question about their performance evaluation.

In order to evaluate an expert system with regard to its performance, it must take into account quality criteria and characteristic measures related to the intelligence of each system itself.

However, software quality criteria can neither be easily measured nor clearly defined. The assessment of a certain software product is relative because the acceptance criteria depend on the context of use, the purpose for which quality characteristics are being described, or even the user.

This is a fuzzy subject because of the diversity of analysis that can be done, including commercial software tools, where the marketing can provide interesting commercial results to a bad or median product and the best ones have bad commercial results.

This is a recurrent problem, as can be seen in academic discussions from some years ago (Plant and Salinas, 1994) that discuss expert system shells. These authors propose the following five-step procedure to create a benchmark system for expert system shells:

1. First, the target area is defined. This is the environment in which the benchmark is to be utilized.
2. The constraints for the benchmark are then laid down—these are the "benchmark goals," which cover aspects such as accuracy, sensitivity, and so on.
3. Having determined these parameters, the techniques by which the benchmark is constructed are selected, along with the statistical approach that will be used to analyze the results of performing the experiments.
4. Having determined the framework in which the benchmark is examined, the standardized benchmark program is designed and implemented.
5. Finally, the benchmark programs are run and the analysis performed.

Cobzaru (2002) presents an interesting analysis of performance quality metrics for expert systems, based on most of the object-oriented software metrics for:

- *Product* (code)—Measurement and evaluation
  - Size estimation
  - Performance level
- *Process*—Measurement and evaluation
  - Behavior simulation
  - Modeling of agent-based systems
  - Agent communication language evaluation
  - Reasonableness of agent derivation
- *Resources*—Measurement and evaluation
  - Middleware evaluation
  - Vendor evaluation
  - Paradigm evaluation

8.8 Usability and System Interfaces

According to Simões-Marques and Nunes (2012), usability covers a broad spectrum of aspects regarding a product. They indicate that usability includes components such as system performance, system functions, user interface, reading materials, language translation, outreach program, ability for customers to modify and extend, installation, field maintenance and serviceability, and advertising or support-group users (Simões-Marques and Nunes, 2012).

The international standard reference on usability is the ISO 9241:1998—Part 11 (*Ergonomic requirements for office work with visual display terminals (VDTs)—Part 11: Guidance on usability*).

ISO/IEC 9126-1:2001 (*Software engineering—Product quality—Part 1: Quality model*) suggests a model based on quality attributes, divided into six main features, and their usability. According to this standard, usability is "the capability of the software product to be understood, learned, used and attractive to the user, when used under specified conditions."

To analyze the usability of a software product, it is necessary to identify who the users are and what their characteristics are, what the user's needs and tasks are, and what the environmental context is (social, organizational, and physical).

According to the referenced authors, another approach is user-centered design, which corresponds to a structured development methodology focused on the needs and characteristics of users. This approach should be applied from the beginning of the development process in order to produce software applications that are more useful and friendly.

According to ISO 13407:1999 (*Human-centered design processes for interactive systems*), there are four key activities related to user-centered design, which should be planned and implemented in order to incorporate the requirements of usability in the process of software development:

1. Understand and specify the context of use.
2. Specify the user and organizational requirements.
3. Produce design solutions.
4. Evaluate design against requirements.

Regarding the specificity of expert system usability, there are some aspects that must be taken into account, namely:

- Principles described in the previous guidelines
- Principles of explanation
  - Make provision for multiple levels of explanation.
  - Allow for ill-posed questions.
- Principles of knowledge acquisition
  - Integrate knowledge acquisition with knowledge application.
- Principles of integration
  - Generate dialogue from knowledge structures.
- Principles of rationality
  - Match the behavior of the system to the cognitive expectations of the user.

8.9 Expert System Example

There are a lot of expert system software applications on the market that anyone can find on the Internet. The following products can be found at http://www.winsite.com/expert/expert+systems/:

- Clipsmm v.0.2.0 Beta
  - Clipsmm was developed with a C++ interface and library for CLIPS.

    CLIPS is an environment for creating rule-based and or object-based expert systems. Clipsmm extends the CLIPS C API in several ways:
    - CLIPS environments are encapsulated in an environment object, as are many of the other CLIPS concepts such as templates (fact templates, not C++ templates), rules, and so on.
    - External functions available to the CLIPS inference engine are simplified using sigc++ slots.
  - This also:
    - Provides the benefit of compiler-type checks on external functions
    - Makes it simple to not only make external functions available, but also external methods of C++ classes
- Optimal Decision System v.3.0
  - Optimal Decision System permits building expert systems in one of the following forms:
    - Decision tree
    - Decision table
  - The Optimal Decision System suite is composed of:
    - Decision Table Designer
    - Decision Tree Designer
    - Code Generator Module and Dictionary Manager
    - Version Manager and Explorer
- JEFF v.1.0.0.beta
  - The Java Explanation Facility Framework (JEFF) can be used with other Java-based (business) rule engines, expert systems, and expert system shells in order to provide an explanation of the inference process.
- Clipsmm—A C++ CLIPS Interface v.0.2.1
  - Clipsmm is a C++ interface to the CLIPS library, a C library for developing expert systems.

The following products are some of many others that can be accessed at http://kbsc.com/rulebase.html:

- Attar
  - XpertRule Builder and Data Mining RBS

- CLIPS
  - C Language Interface Production System
- Drools
  - Dynamic rules object-oriented system, Java RBS; mostly rules in XML
- The Haley Enterprise—CIA
  - High-end C/C++ and Java
- ILOG Rules and JRules
  - High-end C/C++, .NET, and Java business rule management system (BRMS)
- InfoSapient
  - Open-source Java RBS
- Jess
  - Java, CLIPS subset RBS
- Jena2
  - Java, semantic web framework, RBS from HP Labs
- JLog
  - Open-source, ProLog in Java system
- JEOPS
  - Java RBS
- JEOPS on SourceForge
  - Java RBS
- JLisa
  - Open-source, CLIPS-like, Java RBS
- JTP:
  - Java Theorem Prover, open-source, Java RBS
- Mandarex
  - Backward chaining, open-source, Java RBS
- OFBiz
  - Open-source Java RBS
- Pellet
  - OWL and OWL DL RBS, use with Jena or OWL API, from MindSwap
- ROWL
  - RBS in OWL for Jess
- SHOP
  - Hierarchial task network from University of Maryland
- Sweet Rules
  - Semantic web rules from MIT
- TyRuBa
  - Open-source, Java RBS

9

Maintenance 4.0

9.1 Background

The concept of Industry 4.0 corresponds to the Fourth Industrial Revolution, aligned to the current trends of automation and data exchange in manufacturing technologies. It includes cyber-physical systems, the Internet of Things, and also cloud computing.

Industry 4.0 creates what has been called a "smart factory." Within modular, structured smart factories, cyber-physical systems monitor physical processes, create a virtual copy of the physical world, and make decentralized decisions. Over the IoT, cyber-physical systems communicate and cooperate with each other and with humans in real time and through the Internet of Services. Both internal and cross-organizational services are offered and used by participants in the value chain.

There are four main principles in Industry 4.0:

1. Interoperability—The ability of equipment, devices, sensors, and people to connect and communicate among each other via the IoT or the Internet of People
2. Virtual industry—The ability of information systems to create a virtual image of real industry assets and processes through data sensors
3. Maintenance—The ability to support system availability and to help people maintain them, including the use of more recent tools like augmented reality and holography
4. Decentralized decisions—The ability through cybersystems to perform tasks at each level as autonomously as possible

The concept of Maintenance 4.0 ought to be understood as one of the main pillars of Industry 4.0. In fact, condition monitoring tools gain a new dimension with intelligent sensors and the IoT. These new devices permit most equipment to maximize its MTBF and minimize its MTTR, which implies maximizing its availability. Chapter 12 deals with condition monitoring, where some of these aspects are described.

Additionally, new technological tools like artificial vision, mixed reality, augmented reality, and visual and acoustic holography are some of many current tools that help Maintenance 4.0 be strategic in support of Industry 4.0.

9.2 Big Data

Big data describes the enormous volume of data (structured and unstructured).

Big data refers to data sets that are beyond the ability of legacy approaches to analyze and manage with an acceptable level of utility and exceed the capacity of conventional systems to process them.

The importance of big data is related to the potential of information that can be extracted and analyzed to reveal new knowledge and to the optimization of the decision processes.

Big data is a fundamental pillar in Industry 4.0 because the enormous volume of data originating from sensor readings, mainly online, is necessary to monitor the equipment's condition.

Big data has five basic features, the 5 Vs (Figure 9.1):

1. Volume—Characterizes the amount of continuously generated data.
2. Velocity—Data flows continuously, with the aim of being acquired and processed in a short time.
3. Variety—The sources are diverse, and many data items are not structured (videos, mail, comments on social networks, etc.).
4. Veracity—Data collected must be reliable and have quality.
5. Value—Data must produce results for aiding in decision-making.

One very important aspect to be evaluated when using big data is open source platforms. The following platforms and respective links are some of the main ones available on the web:

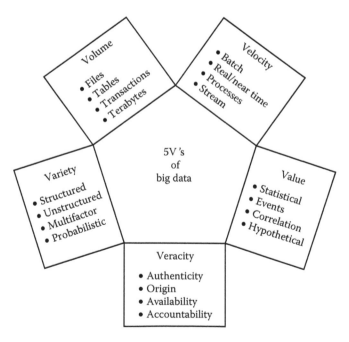

FIGURE 9.1
The 5 Vs of big data.

TABLE 9.1 Processing Framework Comparison—Technical Requirements

	Hadoop	Spark	Storm	Flink	H2O
Processing model	Batch	Batch, streaming	Streaming	Batch, streaming	Batch
Software requirements	Java Development Kit (JDK) 1.7, Secure Shell (SSH)	JDK 1.6	None	Cygwin for Windows, JDK 1.7.x, SSH	JDK 1.7
Programming languages	Java	Java, Python, R, Scala	Any	Java, Scala	Java, Python, R, Scala
Machine learning tools	Mahout	MLLib, Mahout, H2O	Scalable Advanced Massive Online Analysis (SAMOA)	Flink-ML, SAMOA	H2O, Mahout, MLLib

TABLE 9.2

Processing Framework Comparison—Performance

	Hadoop	Spark	Storm	Flink	H2O
Development maturity	Medium	Low	Low	Low	Medium
Modularity	High	High	Low	Medium	Medium
Integration	Low	Medium	Medium	Low	Low

- Apache Hadoop (http://hadoop.apache.org/);
- Apache Spark (http://spark.apache.org/);
- Apache Storm (http://storm.apache.org/);
- Apache Flink (https://flink.apache.org/);
- H2O (www.h2o.ai/).

Tables 9.1 and 9.2 make the comparison among the preceding open source platforms in two ways.

9.3 Internet of Things

The term Internet of Things refers to the networked interconnection of objects that are equipped with ubiquitous intelligence. The IoT integrates every object for interaction via embedded systems, which leads to a highly distributed network of devices communicating among themselves and with humans.

The term IoT refers to things such as devices or sensors that connect, communicate, or transmit data with or between each other through the Internet (Figure 9.2).

The IoT arose due to the exponential growth of the number of sensors and points of signal collecting and transmitting data, which can be, among others, the following:

- Maintenance sensors
- Text messages

FIGURE 9.2
The IoT.

- Videos and images
- Sales data and marketing
- Data based on geolocation

9.4 Sensorization and Data Communications

There are a lot of sensors and other devices that can read and transmit real data to be processed online or later. Some examples of sensors are the following:

- Vibration
- Speed
- Electrical voltage
- Electrical current
- Temperature
- Ultrasound
- Infrasound
- Oil variables
- Chemical
- Effluents

TABLE 9.3

Types of Sensors and Sample Vendors Used in Condition Monitoring

Measurement	Sensor	Frequency Range	Possible Signal Conditioning Needs	Vendors
Vibration	Accelerometer	>100 Hz	IEPE AC/DC coupling ±24 V input or AC couple Antialias filter	IMI Sensors Connection Technology Corporation Endevco/Wilcoxon
Vibration	Velocity	>20 Hz to <2 kHz	IEPE AC/DC coupling ±24 V input or AC couple Antialias filter	IMI Sensors Connection Technology Corporation Endevco/Wilcoxon
Vibration	Proximity probe (displacement)	<300 Hz	Modulator/demodulator Antialias filter ±30 V input range	Connection Technology Corporation
Speed	Proximity probe	<300 Hz	Modulator/demodulator Antialias filter ±30 V input range	Connection Technology Corporation
Speed	Magnetic zero speed	Up to 15 kHz	24 V DC power ±20 V	Honeywell SPECTEC
Motor current	Current shunt current clamp	Up to 50 kHz	±333 mV or ±5 V	Magnelab
Temperature	RTD Thermocouple	Up to 10 Hz	Noise rejection, excitation, cold-junction compensation	NI
Temperature	Infrared camera	Multiple frames per sec	GigE Vision over Ethernet connection	FLIR Systems
Pressure	Dynamic pressure	>100 Hz	AC/DC coupling IEPE (some models) ±24 V or AC coupling Antialiasing filter	Endevco PCB Kulite Kistler
Oil quality Oil particulate	Viscosity Contamination Particulates	Up to 10 Hz	mA current input ±10 V input 50/60 Hz noise rejection	Kittiwake Honeywell HYDAC Poseidon Systems
High-frequency "noise"	Ultrasonic	>20 kHz	AC/DC coupling ±24 V input range Antialiasing filter	UE Systems

Table 9.3 presents some types of sensors and the respective sample vendors used in condition monitoring from National Instruments (http://www.ni.com/white-paper/52461/en/).

Communication can be accomplished in several ways, but it needs to have a well-defined protocol in order for the devices to communicate among themselves. A protocol is a set of rules for a particular type of communication, and its transmission is carried out in a frame format. All communication protocols must always be according to the open systems interconnection (OSI) model, which is divided into seven layers, regardless of the physical medium or type of connection, as Table 9.4 shows.

TABLE 9.4

OSI Model with Seven Layers

N°	Layer	Function
7	Application	Communication aspects
6	Presentation	Data representation
5	Session	Dialog control
4	Transport	Reliability transport
3	Network	Information forwarding
2	Data connection	Errors and flow control
1	Physical	Bit sending and reception

For many years, Ethernet has been the overwhelmingly accepted choice as the local area network (LAN) in companies, having been developed initially for use in offices. This technology is strongly supported by the Institute of Electrical and Electronics Engineers (IEEE) 802.3 standard, making it a viable communication network and very accessible to its users. However, this technology was not prepared to be applied in an industrial environment in its standard format. Because of this, today there are several industrial Ethernet communication protocols that have been developed to establish a data flow among equipment. Some of the most commonly used industrial Ethernet protocols are EtherCAT, Ethernet/IP, Profinet, POWERLINK, and Modbus/TCP (Warren, 2011):

- EtherCAT—Ethernet for Control Automation Technology is a communication protocol of industrial Ethernet real-time technology that was originally developed by Beckhoff Automation. This protocol is described in IEC61158 and is suitable for hard and soft real-time requirements in automation technology, measurement, and testing, among many other applications.
- Ethernet/IP—This is an industrial Ethernet protocol designed by Rockwell Automation and currently managed by the Open DeviceNet Vendor Association (ODVA). This protocol applies the common industrial protocol (CIP) over Ethernet standard and Transmission Control Protocol (TCP)/Internet Protocol (IP) technologies. CIP uses the producer/consumer model instead of the client/server model. The producer/server model decreases network traffic and increases transmission speed. The CIP protocol implements a path for sending messages from system-producing devices to consumer devices, (Lin and Pearson, 2013).
- PROFINET—This is an industrial Ethernet protocol developed with the contributions of Siemens and other members of the Profibus User Organization (PUO). The communication system of this protocol specifies the transfer of data between input/output controllers, parameterization, diagnosis, and implementation of a network. This protocol can be divided into performance classes depending on the time requirements (PROFINET RT for non–real-time or soft–real-time and PROFINET IRT for hard–real-time):
  - PROFINET RT sends a load and data in noncritical real time within a cycle, adopting the producer/consumer principle. A real-time channel is reserved for high-priority loads that are transmitted directly via the Ethernet protocol. However, the configuration of data and diagnostics is sent via the User

Datagram Protocol (UDP)/Internet Protocol (IP) protocol. The 10-millisecond time cycles can be used for input/output applications.

- PROFINET IRT, by applying time division multiplexing based on switch management, provides synchronized clock cycle times below 1 millisecond, which are used in applications, for example, motion control.

- POWERLINK—This is an industrial Ethernet protocol developed by the Austrian automation company Batman and Robin (B&R), characterized by time cycles in the microsecond range, universal applicability, and maximum network configuration flexibility. This protocol provides all the characteristics of the Ethernet standard, including traffic crossing and free choice of network topology. POWERLINK uses a mixture of timeslot and polling procedures to achieve isochronous data transfer. In order to ensure coordination, a Power Line Communication (PLC) is designated to be the managing node (MN). This manager enforces the cycle times, which serve to synchronize all the devices and the cyclic controls of the data communication. The other devices operate as controlled nodes (CNs). In the course of a clock cycle, the MN sends "Poll Request" messages for each CN, which transmits data from the MN to any CN that is connected. Each CN responds immediately to this request with a "Poll Response" that all other elements can hear. The POWERLINK cycle consists of three periods:

  - The first period is known as the "Start Period," in which the MN sends a start-of-cycle (SoC) frame to synchronize all the CN devices;

  - The second period is when the cyclic isochronous data exchange takes place, where the bandwidth is optimized by multiplexing;

  - The third period marks the beginning of the asynchronous phase, which allows the transfer of large non–time-critical data packets.

    This protocol distinguishes between real-time and non–real-time domains. Because the data transfer is in asynchronous periods, it supports IP standard frames, and the routers separate data securely and transparently from real-time data.

- Modbus/TCP is an extension of the Modbus protocol that was developed by Modicon (now a division of Schneider Electric). It implements the same services and same object model as previous versions, using the Ethernet standard for the transfer of data packets via TCP/IP. Messages are supported by networks configured on client/server and peer to peer. All object messages are sent between two elements to execute transaction services, which are associated with a request service and consist of a requested message and the corresponding response message. Typically, Modbus applications operate on a client/server model, where a client scans the server for information. Because of the peer-to-peer capability, Modbus/TCP may support other communication methods, such as asynchronously reporting servers in cyclic or state change mode using the notification service. However, determinism can be compromised when using asynchronous reporting methods.

Regarding industrial wireless, this technology is a relatively new concept in the industrial environment. Initially, there were a lot of doubts about the use of this type of communication due to the preference for data transmission through cables in order to guarantee reliability around noise in hostile environments. However, performing the installation of cable communication networks is an expensive process and does not always allow connection between physical assets that are in remote locations. Nowadays, industrial wireless technology is well accepted for communication with several devices responsible for the processes of various industries. In fact, wireless enables:

- Very high time-critical responses, exceeding cable solutions in some cases
- Eliminates problems that sometimes arise in communications due to faulty cables and connectors
- Allows monitoring and control of equipment via the Internet and cloud

Some of the solutions that can be applied at the industrial level via wireless are the following:

- WirelessHART—This is a wireless sensor network technology based on the highway addressable remote transducer (HART) protocol. Standardized by IEEE 802.15.4, this technology was developed with the purpose of interconnecting the equipment that communicates via HART over wireless networks in industrial processes. WirelessHART stands out for its data security in transmission, high reliability thanks to its routing in mesh architecture, and low power consumption due to its communications in synchronized time.
- Bluetooth—This is wireless network technology based on the IEEE 802.15.1 standard. This protocol was designed for low-power communications and is based on low-cost microchips. This protocol stands out for its extremely reliable transmissions through redundant transmission channels, high coexistence capacity in unfamiliar wireless environments, and parallel operations with several Bluetooth systems at a single point due to the efficient use of frequency. This technology allows interconnection via wireless network with serial communication equipment, Ethernet, and analog/digital sensors.
- ZigBee—This is a communication protocol based on the IEEE 802.5.4 standard, which is an industry-standard wireless network technology that operates at 2.4 GHz. This protocol stands out for its mesh architecture that allows extension of reach to each piece of equipment and provides a more reliable connection due to multiple redundant paths. If any of the paths are interrupted for some reason, another is automatically used.

9.5 Hardware and Software Options

There are several ways to transmit data signals around the world, because commercial network providers, mainly for voice and multimedia, including satellite, can be used for data streaming and specialized uses.

In the ambit of Maintenance 4.0, the challenge is to use the current technology of Industry 4.0, as described in the above sections, making all devices communicate correctly and, finally, to communicate with EAMs/CMMSs with the objective of maximizing equipment availability at the most rational cost.

An intelligent system for early detection of malfunctions or determination of the condition of equipment operation is the key factor for planning interventions and maximizing resources, including financial.

One of the current problems in the integration of technologies, hardware, and software relates to the multiplicity of existing communication protocols of condition sensor devices to communicate with storage and treatment systems—due to the nonstandardization in communications among them. This makes it very difficult, if not impossible, to develop management platforms for widespread use.

To help to solve this type of problem, the MIMOSA open standard protocol/ Operations & Maintenance (O&M) Interoperability aims to help solve the problem of the nonstandardization of communication between devices of "things," with an emphasis on those associated with condition/predictive maintenance.

The MIMOSA protocol aims to standardize the interface between factory floor systems (including predictive maintenance) and EAM systems. The MIMOSA standard is complementary to OPC (formally known as object linking and embedding for process control), which addresses the real-time interface of the devices. There is an umbrella organization called Open O&M, which has a collaboration with MIMOSA, the OPC Foundation, and the ISA SP95 committee.

An important step in the evolution of the MIMOSA protocol was its progression toward a protocol focused on storage for a message-centered protocol. The original version of the MIMOSA protocol was based on a data model called the common relational information schema (CRIS), which was a data model that included database scripts for implementation in SQL Server and Oracle. It created XML Schema Definition (XSD) schemas that were mapped to CRIS schemas, but many vendors remained focused on the protocol storage.

Comparing the protocol to OPC, it became clear that there is a need for a messaging protocol to standardize the interface between factory floor systems and EAM systems. OPC standardized the interface for real-time data retrieval from factory floor devices. What is necessary for EAM systems is a messaging protocol that could do the same for uses such as the automatic generation of working orders, upload of measurement points, retrieval of physical asset data, auditing of the work done, and historical uses. As each EAM supplier had its own database implementation, the interest was primarily in the messaging layer, rather than an additional storage layer. The Tech-eXtensible Markup Language (XML) XSDs and Tech-XML-Services Web service specifications shifted the focus of the messaging layer to integration.

In addition, it can be seen that current solutions can leverage Tech-XML–based Web service communication to link them to EAM systems, but also leverage the CRIS database for standardized reporting capabilities. More and more suppliers are creating reliability analyses and reporting tools based on the CRIS data model, which implies an incremental easing in the linkage to proprietary EAM databases to make reports.

10

Forecasting

10.1 Background

Maintenance planning involves the use of several algorithms, with those based on time series the most common. However, one of the major problems with maintenance activity is the lack of historical data, so the use of complex models that need data with a long history is extremely difficult. From this perspective, models based on moving averages, with short histories and exponential smoothing, are usually the most appropriate.

Being this the main focus of this chapter, at last, it examines several methods that may help with forecasting, namely neural networks, discrete system simulation, and the support vector machine (SVM), among others.

10.2 Time Series Forecasting

Time series forecasting is one classical way to manage historical data to make forecasts. To analyze historical data through a time series, some types of data patterns must be taken into account:

    i. Horizontal—The data values fluctuate around a constant value.

    ii. Trend—There is a long-term increase or decrease in the data.

    iii. Seasonal—A series is influenced by seasonal factors and has a periodic cycle.

    iv. Cyclical—The data have climbs and falls that do not have a fixed period.

In practice, many data series have combinations of the preceding patterns.

There are many techniques to manage time series, such as time plots, autocorrelation, and scatter diagrams. For long and erratic series, or when the historical record is short, as is the case with much equipment, the solution is methods like moving averages and exponential smoothing. Some good references on this subject are Makridakis and Wheelwright (1989) and Makridakis et al. (1997).

10.2.1 Moving Average Method

The moving average method makes a forecast for the next period from the average of n previous periods. The formula for calculating the value of the next period is given by:

$$S_{t+1} = \frac{\sum_{i=t-n}^{i=t} X_i}{n}$$

(10.1)

where:

S_{t+1} = the forecast for the next period

X_i = the actual value recorded for each of the earlier n periods between $i = t - n$ and $i = t$

The following example, Table 10.1 and Figure 10.1, illustrates the application of this method for two ranges of values of three (3M) and five periods (5M), respectively (the letter M means month).

10.2.2 Exponential Smoothing Method

The exponential smoothing method uses an estimate of historical (past) values to make a forecast for the current period. As a consequence, the calculation of the next period value requires only the actual value for the current period and the corresponding forecast value for this period. Additionally, it is necessary to apply a smoothing parameter α that corresponds to the weight history that should be given in the calculation of the value for the next period. The value of this parameter is located in the interval between 0 and 1: $\alpha \in [0,1]$.

The calculation formula for the next period is the following:

$$S_{t+1} = \alpha \cdot x_t + (1-\alpha)S_t \Leftrightarrow S_{t+1} = \alpha \sum_{i=0}^{t} (1-\alpha)^i x_{t-i}$$

(10.2)

TABLE 10.1

Forecast for Three and Five Periods

Period	Observed Value	3M Forecast	5M Forecast
1	1950		
2	1430		
3	1830		
4	1900	1737	
5	2900	1720	
6	1800	2210	2002
7	1675	2200	1972
8	1330	2125	2021
9	2250	1602	1921
10	2620	1752	1991
11	2410	2067	1935
12		2427	2057

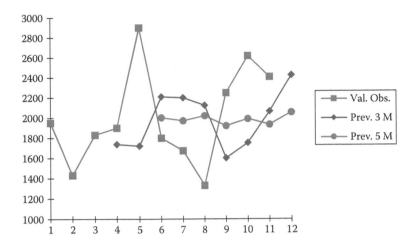

FIGURE 10.1
Graph of observed and predicted values.

with

$$0 \le \alpha \le 1$$

where:
S_{t+1} = the forecast for the next period
x_t = the actual value recorded at the present time
S_t = the forecast value for the present time
α = the smoothing parameter

The following example, Table 10.2 and Figure 10.2, illustrates the application of this method for two values of the smoothing parameter.

TABLE 10.2

Forecast for Two Values of the Smoothing Parameter

Period	Observed Value	Forecast $\alpha = 0.1$	Forecast $\alpha = 0.9$
1	1950		
2	1430	1950	1950
3	1830	1898	1482
4	1900	1891	1795
5	2900	1892	1890
6	1800	1993	2799
7	1675	1974	1900
8	1330	1944	1697
9	2250	1882	1367
10	2620	1919	2162
11	2410	1989	2574
12		2031	2426

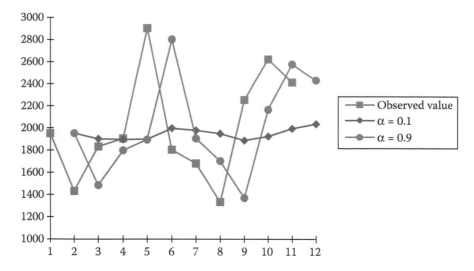

FIGURE 10.2
Graph of observed and forecast values.

10.2.3 Second-Order Exponential Smoothing Method

When there is a tendency to climb or fall, the second-order exponential smoothing method should be used.

The second-order exponential smoothing method corresponds to what can be called double smoothing. Formally, it is the previous model, with the difference that the second order is applied to the prediction obtained by the exponential smoothing method (first order).

The advantage of the second-order exponential smoothing method from the previous one is that the difference between the actual value and that of the first-order forecast is approximately equal to the difference between the last and the prediction of the second order. In this way, the error between the forecast of the first order and the actual value can be corrected. This forecast provides a third estimate that, when some stability and tendencies occur, is closer to the actual value.

The formulas for calculation of the next period are given by:

$$S'_t = \alpha \cdot x_t + (1-\alpha)S'_{t-1} \tag{10.3}$$

$$S''_t = \alpha \cdot S'_t + (1-\alpha)S''_{t-1} \tag{10.4}$$

$$0 \leq \alpha \leq 1$$

$$a_t = 2 \cdot S'_t - S''_t \tag{10.5}$$

$$b_t = \frac{\alpha}{1-\alpha}(S'_t - S''_t) \tag{10.6}$$

$$S_{t+m} = a_t + b_t \cdot m \tag{10.7}$$

where:

S'_t = first prediction of the second order for the next period

S''_t = second prediction of the second order for the next period

α = smoothing parameter

a_t = coefficient that supports the calculation of the second-order forecast

b_t = coefficient that supports the calculation of the second-order forecast—second smoothing

S_{t+m} = the forecast value of the second order for the period m

The following example, Table 10.3 and Figure 10.3, illustrates the application of the method by calculating the value forecast for the next period.

TABLE 10.3

Forecast for Second-Order Exponential Smoothing

Period	Observed Value	S'	S''	a	b	S
1	1950	1950	1950			
2	2010	1950	1950	1950	0	1950
3	2030	1956	1951	1961	1	1962
4	2035	1963	1952	1975	1	1976
5	2045	1971	1954	1987	2	1989
6	2053	1978	1956	2000	2	2002
7	2000	1986	1959	2012	3	2015
8	2070	1987	1962	2012	3	2015
9	2073	1995	1965	2025	3	2029
10	2090	2003	1969	2037	4	2041
11	2082	2012	1973	2050	4	2054
12		2019	1978	2060	5	2064

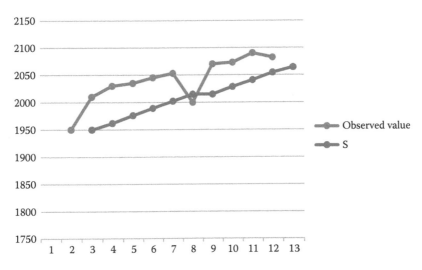

FIGURE 10.3

Chart of observed values and second-order exponential smoothing forecast.

10.2.4 Nonperiodic Exponential Smoothing Method

Above, it was noted that one of the difficulties in maintenance planning is the historical absence of data for many physical assets. Furthermore, when new equipment is put into operation, it does not have any historical records, so it is not possible to make predictions based on past data. Additionally, there are many assets for which it is not possible to periodically register the historical data. As a consequence, it becomes imperative to use a method to solve these difficulties.

The method of simple exponential smoothing responds to these problems, but with regular, periodic intervals, and only the last observed values and the forecast value for the period of this observation are necessary to forecast the next value. It is based on this method that a prediction algorithm was developed that may be used to plan maintenance at aperiodic intervals. This method allows responses to variables such as operating hours, kilometers, or others that are not registered regularly.

As seen in the preceding section, from the exponential smoothing method, it is known that, given the last parameter value x_t and the estimate for that period S_t, the forecast value for the next period is given by Equation 10.2.

When the time intervals are not equal, the previous method cannot be applied directly. To solve this problem, an intermediate variable is defined corresponding to the regular interval time series, which will be the reference for the nonregular sampling values. For example, for a device that requires planned interventions every 5000 hours of operation, the average value of the monthly hours of operation may be evaluated (e.g., 500 hours), from which it is possible to determine the data of the next intervention. This intermediate variable can assume any value since it is below the lowest actual reading range (Farinha, 1994). The calculation method is as follows:

$$m_i = \frac{D_{i-1}}{I_{i-1}} + (1-\alpha)^{I_i-1}\left(m_{i-1} - \frac{D_{i-1}}{I_{i-1}}\right) \tag{10.8}$$

$$v_i = I_{i-1}\left(\frac{D_{i-1}}{I_{i-1}} - m_{i-1}\right)^2 + (1-\alpha)^{I_i-1}\left(v_{i-1} - I_{i-1}\left(\frac{D_{i-1}}{I_{i-1}} - m_{i-1}\right)^2\right) \tag{10.9}$$

$$I_i = \frac{I_{nc}}{m_i} \tag{10.10}$$

where:
D_{i-1}—Total evolution of control variable—previous interval
I_{i-1}—Interval during which D_{i-1} occurred—previous interval
α—Smoothing coefficient
m_{i-1}—Average variation of the control variable—previous interval
v_i—Variance of m_i—current interval
v_{i-1}—Variance of m_i—previous interval
I_i—Next interval
I_{nc}—Increment of control variable
m_i—Average variation of the control variable—current interval

In the case of normal distribution, a confidence interval of 95% can be determined for the predicted value using the following formula:

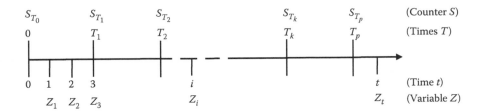

FIGURE 10.4
Nonperiodic time series.

$$I_i = \left| \frac{I_{nc}}{m_i} - 1.65 \sqrt{\left| I_{nc} \frac{v_i}{m_i^3} \right|} \right|$$ (10.11)

Figure 10.4 illustrates the relation between the periodic intervals Z_i and aperiodic readings of control variable T_i, which consequently gives rise to aperiodic interventions in terms of calendar time, despite its origin of periodic control variables.

From the above points, formulas of exponential smoothing may be applied and then extended to nonperiodic intervals, as follows:

The value of the control variable (CV) for the next time is given by:

$$M_{T_p} = x_{T_p} + (1-a)^{I_{T_p}} \left(M_{T_{p-1}} - x_{T_p} \right)$$ (10.12)

$$0 < a < 1 \quad e \quad x_{T_p} = \frac{S'_{T_p}}{I_{T_p}}$$

$$S'_{T_p} = S_{T_p} - S_{T_{p-1}}$$ (10.13)

$$I_{T_p} = T_p - T_{p-1}$$

$$E\left[S_{T_{p+1}}\right] = S_{T_p} + I_{T_{p+1}} \cdot M_{T_p}$$ (10.14)

The variance value for the next time is given by:

$$U_{T_p} = g_{T_p} + (1-b)^{I_{T_p}} \left(U_{T_{p-1}} - g_{T_p} \right)$$ (10.15)

$$g_{T_p} = I_{T_p} \cdot \left(x_{T_p} - M_{T_{p-1}} \right)^2$$ (10.16)

$$E\left[g_{T_{p+1}}\right] = I_{T_{p+1}} \cdot U_{T_p}$$ (10.17)

where:
  a and b—Smoothing coefficients
  M_{T_p}—Average forecast of CV to the next interval
  $M_{T_{p-1}}$—Average forecast of CV for the previous interval
  x_{T_p}—CV average in the current interval
  I_{T_p}—Current interval width (time)
  $I_{T_{p+1}}$—Width of the next interval (time)
  S_{T_p}—Forecast value of the CV to the current time

$S_{T_{p-1}}$—Forecast value of the CV to the previous time
S'_{T_p}—Value difference of CV between the present and previous times
U_{T_p}—Expected variance for the current time
$U_{T_{p-1}}$—Expected variance to the previous time
g_{T_p}—Variance of the average for current time
$g_{T_{p+1}}$—Variance of the average to the next time

To clarify the new calculation method for the CV of nonperiodic intervals, the following example is presented:

- A car maintenance workshop established a planned maintenance policy for its vehicles, despite their use being irregular. For this purpose, it used nonperiodic exponential smoothing for the vehicles whose maintenance intervals were 10,000 km. Knowing that the average use is 500 km per week, with a standard deviation of 200 km, which is the date of the next intervention?

The resolution of this planning problem using the nonperiodic exponential smoothing method is as follows:

$$a = b = 0.1; \quad S'_{T_0} = 10,000\,\text{km}; \quad M_{T_0} = x_{T_0} = 500\,\text{km}; \quad g_{T_0} = 200\,\text{km}$$

$$I_{T_0} = \frac{10,000}{500} = 20 \text{ weeks}$$

But, the maintenance intervention was done at 17 weeks and registered the value of 7800 km. When is the date of the next intervention?

$$I_{T_1} = 17 \text{ weeks}$$

$$S'_{T_1} = 7800\,\text{km}$$

$$x_{T_1} = \frac{7800}{17} \cong 459\,\text{km}$$

$$M_{T_1} = 459 + (1-0.1)^{17}(500-459) \cong 466\,\text{km}$$

$$I_{T_2} = \frac{10,000}{466} \cong 21.5 \text{ weeks}$$

$$E[S_{T_2}] = S_{T_1} + I_{T_2} \cdot M_{T_1} = 7800 + 21.5 * 466 = 17,819\,\text{km}$$

Calculation of the variance:

$$U_{T_1} = g_{T_1} + (1-b)^{I_{T_1}}(U_{T_0} - g_{T_1}); \quad U_{T_0} = 200^2 = 40,000$$

$$g_{T_1} = I_{T_1} \cdot (x_{T_1} - M_{T_0})^2 = 17(459-500)^2 = 28,577 \cong 169^2$$

$$U_{T_1} = 28,577 + (1-0.1)^{17}(40,000 - 28,577) \cong 30,482 \cong 175^2$$

$$E[g_{T_2}] = I_{T_2} \cdot U_{T_1} = 21.5 * 30,482 \cong 810^2$$

TABLE 10.4

Map for Planning through Nonperiodic Exponential Smoothing

Period	Forecast CV	Real CV	Forecast Time	Real Time	Forecast Average	Real Average	Total Counter	Forecast Variance	Real Variance	Forecast Standard Deviation	Real Standard Deviation
1	10,000	7800	20	17	500	459	10,000	40,000	28,824	200	170
2	10,000	11,300	21	18	466	628	17,800	30,687	472,901	175	688
3	10,000	10,100	17	21	603	481	29,100	406,527	315,116	638	561
4	10,000	12,000	20	22	494	545	39,200	325,118	57,444	570	240
5	10,000	13,000	19	16	540	813	51,200	83,803	1,184,419	289	1088
6	10,000	8400	13	17	762	494	64,200	980,472	1,220,697	990	1105
7	10,000	9500	19	21	539	452	72,600	1,180,634	156,858	1087	396
8	10,000	9000	22	20	462	450	82,100	268,879	2803	519	53
9	10,000	10,000	22	19	451	526	91,100	35,151	106,524	187	326
10	10,000	8900	19	21	516	424	101,100	96,882	179,260	311	423
11	10,000	11,500	23	18	434	639	110,000	170,246	756,228	413	870
12	10,000	12,100	16	19	608	637	121,500	668,276	15,670	817	125
13	10,000		16		633		133,600	103,827		322	

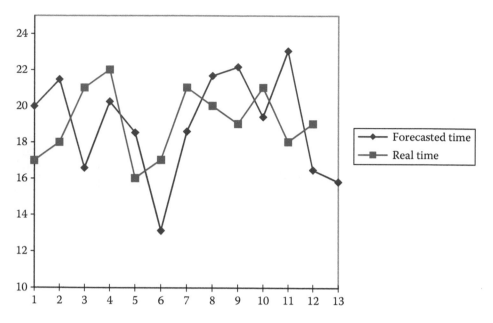

FIGURE 10.5
Graph of forecast and actual times.

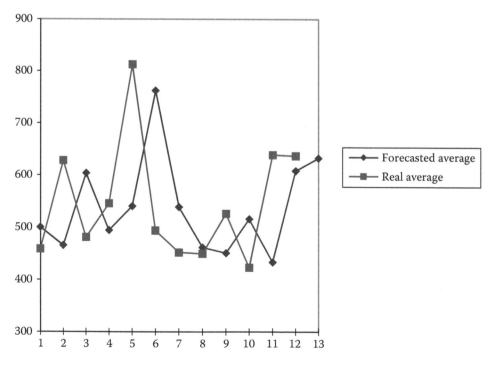

FIGURE 10.6
Graph of the forecast and actual average.

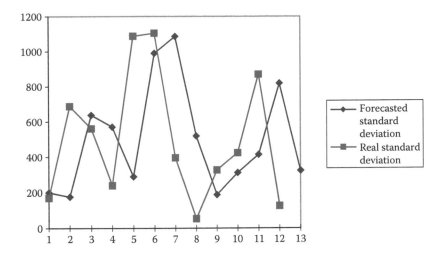

FIGURE 10.7
Graph of forecast and real standard deviation.

Table 10.4 and Figures 10.5 through 10.7 illustrate the application of the method in this problem.

10.3 Neural Networks

Neural networks (NNs) are computational models used in machine learning, computer science, and other disciplines, based on a large collection of simple connected units called artificial neurons, considered analogous to axons of a human brain. The connections among neurons carry an activation signal of variable strength. These systems can be trained from examples and are applied in areas where the solutions are difficult to express using a traditional computer program. In a way similar to other machine learning methods, neural networks have been used to solve a wide variety of tasks that are difficult to solve using ordinary traditional programming.

It is because of the preceding reasons that NNs are included in this chapter, namely for situations of forecasting the next interventions in condition monitoring with predictions. The prediction of the next intervention based on one or more condition variables requires some algorithms, usually parameterized according to the specificity of each piece of equipment. These algorithms, which are deterministic or stochastic, do not have any learning capacity, in contrast to tools like neural networks.

The model of a time series generated by a collection of sources can be used not only in classification problems, but also for prediction, namely to forecast the next value, as seen in previous sections of this chapter.

A modular neural network can be considered a network composed of subnetworks or modules in which each module may be specialized in a particular task. It is hoped that the combination of modules will yield superior performance, greater noise robustness, a shorter training cycle, and so on (Petridis and Kehagias, 1998).

A neural network can approximate any continuous function, as is the case with the classical methods used for time series prediction like the auto regressive moving average (ARMA) and auto regressive integrated moving average (ARIMA) methods, which assume that there is a linear relationship between inputs and outputs. Neural networks also have the advantage of approximating any nonlinear functions without any historical information about the properties of the data series.

10.4 Discrete System Simulation

Discrete event simulation models the operation of a system as a discrete sequence of events in time. Each event occurs at a particular instant in time and marks a change of state in the system. Between consecutive events, no change in the system may occur. Thus, the simulation can directly jump in time from one event to the next.

Some basic applications of simulation correspond to the case discussed in this chapter, namely maintenance planning and forecasting.

The simulation usually has the following characteristics:

- Uses a system definition to run in time-based functioning
- Includes random variables
- Can be a continuous or discrete time event

The components of a discrete event simulation are the following:

- Activities—Things happen to entities at some time, which may have a probability distribution
- Queues—The entities wait an undetermined time
- Entities—Entities wait in queues or get sequenced in their activities
  - Entities can have attributes like kind, weight, date, priority

A discrete system simulation process must include the following characteristics:

- Predetermined starting and ending points
- A method of keeping track of the time elapsed since the process began
- A list of discrete events that have occurred since the process began
- A list of expected discrete events until the process is expected to end
- A tabular record of the function for which the simulation is engaged

In fact, for simulation discrete systems, namely the ones discussed in this book, that go with asset management and maintenance planning and forecasting, simulation can be a useful tool.

A model can be used to investigate a wide variety of questions about system behavior. Potential new states of the system can be simulated in order to predict their impact on system performance. Simulation modeling can be used both as an analysis tool for predicting new values of new states in existing systems and as a design tool to predict the performance of new systems (Banks, 2009).

10.5 Support Vector Machines

Support vector machines are based on the concept of decision planes that define decision boundaries.

A decision plane is one that separates a set of objects having different class memberships.

SVM is primarily a classifier method that performs classification tasks by constructing hyperplanes in a multidimensional space that separates cases of different class labels.

SVM supports both regression and classification tasks and can handle multiple continuous and categorical variables.

To construct an optimal hyperplane, SVM employs an iterative training algorithm, which is used to minimize an error function. According to the form of the error function, SVM models can be structured into four distinct groups:

1. Classification SVM Type 1
2. Classification SVM Type 2
3. Regression SVM Type 1
4. Regression SVM Type 2

The prediction of time series is a very difficult task, as noted in previous sections: it involves forecasting the behavior of a complex system based on simplistic data points along the time axis. Additionally, it can encounter the problem of nonperiodicity, which implies an additional difficulty, solved as proposed in a section above.

The data can only be treated as stochastic in nature. However, any *a-priori* structure of the data cannot be assumed.

SVM methodology provides good performance for nonlinear problems and does not require prior knowledge of the structure of the data, which is more than adequate for historical asset maintenance. This makes the SVM a good tool for time series prediction.

One of the main problems of time series analysis, the forecasting of time series, can be very easily stated as a pure numerical problem. Other learning tasks, such as classification or similarity computation of time series, can also be formulated as purely numerical problems. Support vector machines can be successfully applied for these kinds of learning tasks.

10.6 Other Prediction Techniques

Time series analysis and prediction are important tools for system modeling. The method of support vector regression (SVR) permits solving prediction problems of complex time series.

The traditional tools for time series prediction may use statistics and neural networks, as described in previous sections. However, statistical tools may not fit complex time series and aren't possible to use when equipment doesn't have historical data of interventions.

Neural networks fit well with nonlinear approximation, but they may not be adequate if excessive training is necessary.

SVM, as described in a previous section, suggests a good tradeoff between the complexity of the model and learning ability to obtain some generalization. But, if the SVM is applied for the problem of regression, it is called support vector regression and is a good tool for time series prediction of complicated dynamic systems.

Another prediction tool is series hazard modeling, which works well with discrete data, as is the case with asset maintenance interventions. These happen once each time and it is possible to know the exact date of all events so that the dependent variable can be calculated as the duration until the next event, which are requisites for series hazard modeling.

Some other tools can be used for forecasting, like artificial intelligence, Markov models, and other methodologies that are also referred to in other chapters of this book. However, their detailed analysis is out of the scope of this book.

10.7 A Case Study

In addition to the examples given in the previous sections, a final example of forecasting based on time series of data for buses run by a company in the road transport passenger sector will be given.

This example considers one effluent, the soot found in the oil, of a specific bus (N° 287), and uses the exponential smoothing method (Equation 10.2) to predict the next intervention. The control variable (Period) used was the kilometer, and the variable soot units used was the percentage (%). Table 10.5 and Figure 10.8 show the evolution of the soot degradation

TABLE 10.5

Application of Exponential Smoothing to Soot (%)

	Soot (%)			
Period	Observed Value	Prediction with $\alpha = 0.1$	Prediction with $\alpha = 0.5$	Prediction with $\alpha = 0.9$
2,451	1.1			
5,214	1.5	1.10	1.10	1.10
9,832	2.7	1.14	1.30	1.46
10,000	3.1	1.30	2.00	2.58
10,000	3	1.48	2.55	3.05
10,000	2	1.63	2.78	3.00
10,000	2.3	1.67	2.39	2.10
10,370	3.3	1.73	2.34	2.28
11,542	3.8	1.89	2.82	3.20
14,000	5.5	2.08	3.31	3.74
15,000	0.8	2.42	4.41	5.32
15,000	2.9	2.26	2.60	1.25
15,000	2.5	2.32	2.75	2.74
17,212	2	2.34	2.63	2.52
20,000	2.5	2.31	2.31	2.05
22,183	1.1	2.33	2.41	2.46

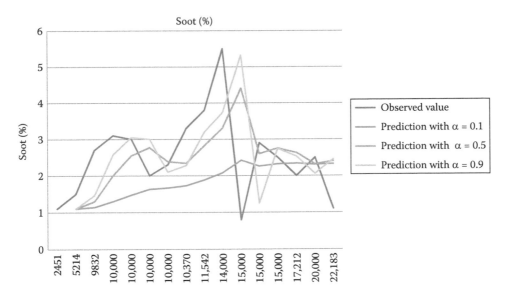

FIGURE 10.8
Graph of exponential smoothing of soot (%).

in the oil tested. The level of reference is 1.5 (danger > 1.5) for a diesel engine. When the variable exceeds this value, the oil must be replaced immediately because the equipment is reaching a high risk level.

Exponential smoothing uses the predicted value for the current period as the historical estimate.

11

Maintenance Logistics

11.1 Background

Problems with logistics in maintenance happen in several situations, but usually they are missed. There are several situations where maintenance logistics have a potential impact on maintenance performance and its costs, namely the following:

- When the facilities and equipment (physical assets) are geographically dispersed
- When the facilities and equipment even located in a single plant that is very large, and the paths are long
- When the physical asset layout is complex and the maintenance paths are longer than expected or not rationalized
- When the spare parts warehouse is far from the physical assets location, or the paths among them are long

As maintenance management becomes more and more rigorous, all of these types of variables must be more closely evaluated and, as a consequence, the subject discussed in this chapter becomes more relevant.

The sections of this chapter present topics ranging from the more structural questions associated with warehouse management, including automatic systems and codification systems, to some of the more useful algorithms about route optimization that can support resource management optimization.

11.2 Warehouse Management Systems and Inventories

Warehouse systems can be conventional, static, or automatic. There are no unique solutions, because all may correspond to the most adequate solution for each specific company considered.

Chapter 16 discusses some optimization methods for maintenance management, including a final example of a traditional warehouse. Visual management, 5S organization, and Lean management are only some of the various tools that can be used to manage according to the current state of the art. Chapter 6, namely Section 6.3, discusses a method for spare parts management.

However, automatic warehouses are becoming more and more a solution that can be adopted for several reasons, particularly because of a lack of space and or to access parts more quickly.

Additionally, robots are interacting more with humans and, in this area, they have enormous potential to help technicians, such as when they need parts that were not planned and that require the technicians to go to the warehouse several times to pick up parts—these tasks may now be done by autonomous robots.

11.3 Basic Identification Tools

Physical assets and spare parts may have a code in order to identify them both in the information system and the material themselves.

Several solutions may be used, such as structured alphanumeric codes, barcodes, and radio-frequency identification (RFID), to name a few.

Regarding structured alphanumeric codes, those used during the last decades have been, over time, replaced by the second ones referred to above. However, their replacement takes a significant amount of time and, because of that, there are many companies that continue with this solution.

Code structure is implemented by dividing the items into families, subfamilies, and so on, down to the level of subdivision necessary to classify all items.

The other types of codification may use a structured codification or not. However, the use of blind codifications is common.

Regarding barcodes, the most common codes used are the following:

- EAN 10 (European Article Number 10)
- EAN 13 (European Article Number 13)

These codes usually have validation by a check digit (CD), and the most commonly used CD is module 11. This validation has as big advantage for validating the code itself before it is entered into the database of the information system.

RFID codes can be active or passive, with the latter the most common.

RFID uses electromagnetic fields to identify and track tags attached to parts and equipment. The tags contain electronically stored information, namely the material code. Passive tags collect energy from a nearby RFID reader's device. Active tags have a local power source, such as a battery, and may operate in a higher space than the first ones. An RFID tag may be embedded in the tracked object.

RFID devices serve the same purpose as a barcode, providing a unique identifier for each object. However, a significant advantage of RFID devices is that this device does not need to be positioned precisely in front of the scanner.

But, RFID solutions also have problems, like reader and tag collision: reader collisions occur when the signals from two or more readers overlap—the tag is unable to respond to simultaneous queries. Systems must be carefully designed to avoid this problem. Tag collisions occur when many tags are present in a small area.

11.4 Transport Systems

Transport systems in maintenance logistics are strategic in most companies, some examples being manual tool carts, automated guided vehicles (AGVs), autonomous robots, vans, or others.

AGVs are driverless, computer-controlled vehicles that are programmed to transport materials through designated pickup and delivery routines on a shop floor. In the maintenance case, they may deliver parts between the warehouse and the technician workplace.

AGVs are typically used where repetitive movements of materials are required, but where almost no human decision-making is required to perform the movements. They are especially useful in serving processes where there are repetitive movements and no barriers.

Autonomous robots are self reliant, having the ability to move themselves throughout the operation without human assistance. They are able to avoid situations that may be harmful to themselves, people, or property. Autonomous robots are also likely to adapt themselves to changing surroundings.

Vans and other types of vehicles are most usual when is necessary to transport people and material long distances, such as among the company's facilities and/or equipment when the plant has a large area, as is the case for some oil and gas plants, wind farms, and similar others.

11.5 Route Planning

Route planning is an extremely complex problem for maintenance applications because of the high diversity of situations that can occur.

The standard route-planning algorithms usually generate a minimum-cost route based on a predetermined cost function. These tools will be discussed in detail in the next section and subsections.

The combination of a global positioning system (GPS) and a geographic information system (GIS) provides great insight into the route choice decision-making process.

GIS is frequently confused with GPS because it is a more generic acronym used to describe a more complex mapping technology that is connected to a particular database. Because it is generic, it represents a broader concept than GPS in its technical sense. GIS is a computer application that is utilized to view and handle data about geographic locations and spatial correlations, among others.

GISs are widely used to optimize maintenance schedules and fleet movements. Typical implementations may result in savings from 10 to 30% in operational expenses (OPEX) through reduction in fuel and staff time. Additionally, they help improve customer service.

A GPS is a network that permits the location of places on earth, whereas a GIS is a computer program that processes data linked to certain places or locations. GIS is a more generic framework compared to the specific GPS network. However, they ought to work together.

In fact, GISs permit the use of algorithms, like those discussed in the next sections, and can define optimized routes in order to minimize maintenance travel costs, including fuel and human costs, among others.

11.6 Tools to Aid Logistics

There are several tools to aid logistics, including maintenance logistics. Some of them are discussed in the next subsections, which are the following:

- Linear programming
  - Linear programming, or linear optimization, is a method that permits to achieve the best outcome through a mathematical algorithm whose requirements are represented by linear relationships. It is a technique that permits optimization of a linear objective function subject to linear equality and inequality constraints.
- Integer programming
  - Integer programming can be described within the problems subject to a mathematical optimization in which some or all of the variables are restricted to integers. Sometimes this concept refers to integer linear programming, in which the objective function and the constraints (other than the integer constraints) are linear.
- Dynamic programming
  - Dynamic programming is an algorithmic technique that is usually based on a recurrent formula and one (or some) starting states. A subsolution of the problem is constructed from previously found ones. Dynamic programming solutions have a polynomial complexity that ensures a much faster running time than other techniques like backtracking, brute-force, and so on.
- Stochastic programming
  - Stochastic programs are mathematical programs where some of the data incorporated into the objective or constraints are uncertain. The outcomes are generally described in terms of elements a of a set A. A can be, for example, the set of the possible maintenance interventions over the next few weeks.
- Nonlinear programming
  - Nonlinear programming involves minimizing or maximizing a nonlinear objective function subject to bound, linear, or nonlinear constraints, where the constraints can be inequalities or equalities. Some example of problems in the maintenance field include reliability improvement and computing optimal trajectories for maintenance planning.
- Queue management
  - A queue management system is used to control queues. This is a typical situation of maintenance planning with many working orders queued to be implemented in the equipment for which they were programmed.
- Ant algorithm
  - The ant (colony) algorithm is used for finding the optimal paths that are based on the behavior of ants searching for food. This algorithm permits, for example, finding the most adequate trip for maintenance that needs to make interventions in dispersed facilities, like wind farms, among others.

- Dijkstra's algorithm
  - Dijkstra's algorithm permits finding the shortest path between nodes in a graph, which may represent, for example, road networks. It represents an interesting tool for maintenance route optimization, especially in conjunction with the ant algorithm.

11.6.1 Linear Programming

Linear programming is a tool to reach the best way to allocate limited resources.

In the case where maintenance must manage limited resources, namely human and material resources, it is constantly necessary to decide the best quantity of each resource to use to execute the planned working hours. The problem is that there are limitations in the available human resources and the fact that the equipment is available for maintenance only a certain number of hours a day. How should the maintenance manager allocate the human resources, materials, and equipment time to execute the required working orders at minimal cost?

As noted before, linear programming, or linear optimization, is a method that permits achievement of the best outcome through a mathematical algorithm whose requirements are represented by linear relationships. It is a technique that permits optimization of a linear objective function subject to a linear equality and inequality constraints.

In linear programming, one of the most useful tool is the simplex method, which has three main strands that it can solve:

1. What it is intended to maximize or minimize, called the objective function, will always be of the form,

$$u = c_1 x_1 + c_2 x_2 + c_3 x_3 + \cdots + c_n x_n + d \tag{11.1}$$

where d is a constant.

2. The main constraints are of the form

$$a_{i1} x_1 + a_{i2} x_2 + a_{i3} x_3 + \cdots + a_{in} x_n \begin{cases} \leq b_i \\ = b_i \\ \geq b_i \end{cases} \tag{11.2}$$

3. The separately stated nonnegative constraints require that all variables be nonnegative.

If we have a linear program with m main constraints and n variables, where $m \leq n$, then it can be said that the linear program is in the *canonical form* if all of the main constraints are equations. From each main constraint, one variable can be picked out that occurs only in that equation and has the coefficient 1. These *distinguished* variables are called *basic*. The remaining variables are called *nonbasic*.

A linear program in canonical form is said to be in *perfect canonical form* if:

1. The objective function is to be maximized.
2. The b_j on the right-hand side of the equations are nonnegative.
3. The objective function is expressed in terms of the nonbasic variables only.

One of the main focuses of this chapter is maintenance logistics, with emphasis on when transports are relevant. This situation happens, for example, when it is necessary to transport resources from m origins to n destinations. The origins are usually warehouses, while the destinations are the different facilities and equipment. The objective is to reach the minimum total cost. The known data of the problem must specify the route networks over which the resources may be transported and the unit cost of transportation for each route.

There are many references about this subject, for example, Sultan (2011) and Paris (2016).

11.6.2 Integer Programming

The linear programming models are continuous, in the sense that decision variables are allowed to be fractional—however, as fractional solutions are not realistic, the following optimization problem must be considered:

$$\text{Maximize} \quad \sum_{j=1}^{n} c_j x_j, \text{with}$$

$$\sum_{j=1}^{n} a_{ij} x_j = b_i, \quad (i = 1, 2, \ldots, m) \tag{11.3}$$

where

$$x_j \geq 0 \quad (j = 1, 2, \ldots, n)$$

x_i is an integer, for some or all $j = 1, 2, \ldots, n$.

As mentioned before, integer programming can be described within the problems subject to a mathematical optimization in which some or all of the variables are restricted to integers.

It is said to be a *mixed* integer program when some, but not all, of the variables are restricted to integers, and it is called a *pure* integer program when all decision variables must be integers.

Again, this approach is adequate for some maintenance logistics problems, namely to minimize the length of the routes both for planned and/or nonplanned maintenance when the set of the n points to be visited is known and the team passes through each point exactly once.

It is necessary to evaluate each specific program to decide on using integer programs, taking into account the following:

- Advantages of restricting variables to integer values
  - More realistic
  - More flexible
- Disadvantages
  - More difficult to model
  - More difficult to solve

One typical problem that can be solved by integer programming is the warehouse location, including the spare parts, that is so strategic in maintenance management.

Another common problem that can be solved through this tool is scheduling, including sequencing and routing, which are inherently integer programs.

There are many references available on this subject, for example, Schrijver (1998), Wolsey (1998), and Conforti et al. (2014).

11.6.3 Dynamic Programming

Dynamic programming is a method that permits solving problems through solutions based on successively solving similar but smaller problems. This approach is used in algorithmic tasks in which the solution of a bigger problem is relatively easy to find if there are solutions for its subproblems. Dynamic programming is an algorithmic technique that is usually based on a recurrent formula and one or some of the starting states. A subsolution of the problem is constructed from previously found ones. Dynamic programming permits the transformation of a complex problem into a sequence of simpler problems. Its essential characteristic is the multistage nature of the optimization procedure.

In dynamic programming, the solutions of each of the subproblems are stored, that is, memorized. The next time the same subproblem occurs, instead of searching for its solution again, the method simply looks up the previously computed solution, thereby saving calculation time.

Dynamic programming algorithms are used for optimization—a dynamic programming algorithm examines the previously solved subproblems and combines their solutions to give the best solution for the problem under analysis.

The main steps to model a problem using a dynamic programming approach are the following:

1. Define the subproblems.
2. Write down the recurrence that relates the subproblems.
3. Recognize and solve the base cases.

When using the dynamic programming approach, a lot of other techniques may be used to help solve the problems, like Markov chains or the ant and Dijkstra's algorithms, among others.

There are many references available on this subject, for example, Bellman (2003), Ben-Daya et al. (2009), and Ulmer (2017).

11.6.4 Stochastic Programming

Stochastic programming is an approach for modeling optimization problems that involve uncertainty. Whereas deterministic optimization problems are expressed by well-known parameters, real-world problems almost always include variables whose values are unknown at the time some action must be taken.

Stochastic programming models are similar to the preceding, but try to take advantage of the fact the probability distributions that model the data are known or can be estimated. Often, these models apply to a data setting for which the decisions are made repeatedly under similar circumstances. The final objective is to find a decision that will perform well on average.

An example is the design of maintenance routes to accomplish maintenance plans in dispersed facilities and equipment of a company. In this case, the probability distributions can be estimated from historical data. The goal is to reach the most rational costs, guaranteeing the maximum facility and equipment availability.

It is usual to divide stochastic programming approaches into two solution methods:

- For problems with a single time period—single-stage problems
  - Single-stage problems try to find a single, optimal decision, such as the best set of parameters for a statistical model given data. Single-stage problems are usually solved with modified deterministic optimization methods.
- For problems with multiple time periods—multistage problems
  - Multistage problems try to find an optimal sequence of decisions, such as scheduling water releases from hydroelectric plants over a two-year period. Because of the dependence on future random variable behavior, direct modification of the deterministic methods creates difficulties in multistage problems. These methods are dependent on statistical approaches and assumptions about problem modeling, such as finite decisions and outcome spaces or a Markovian model to represent the decision process.

There are many references available on this subject, for example Birge and Louveaux (2011) and Herrera (2017).

11.6.5 Nonlinear Programming

As referred to above, nonlinear programming involves minimizing or maximizing a nonlinear objective function subject to bound, linear, or nonlinear constraints, where the constraints can be inequalities or equalities.

A general optimization for a nonlinear programming problem starts by selecting n decision variables $x_1, x_2, \ldots, x_n$ from a given possible region aiming to optimize (minimize or maximize) a given objective function of the decision variables:

$$f(x_1, x_2, \ldots, x_n) \tag{11.4}$$

The problem is called a *nonlinear programming problem* if the objective function is nonlinear and or the possible region is determined by nonlinear constraints. Thus, for the maximization approach, the general nonlinear program is modeled by:

$$\text{Maximize } f(x_1, x_2, \ldots, x_n), \tag{11.5}$$

Subject to:

$$g_1(x_1, x_2, \ldots, x_n) \leq b_1$$
$$\vdots \qquad \vdots \tag{11.6}$$
$$g_m(x_1, x_2, \ldots, x_n) \leq b_m$$

And each one of the constraint functions, from g_1 to g_m, is given.

There are many references available on this subject, for example, Bazaraa et al. (2006) and Bertsekas (2016).

11.6.6 Queue Management

As stated before, the queue management is a typical situation of maintenance planning, with many working orders queued to be executed in the facilities and equipment for which they were programmed. Another maintenance area where this tool may also be applied is in spare parts management.

The first traditional queuing discipline is called first-in first-out (FIFO). But this is not the only one possible. For example, in stock management, it means that the oldest items are recorded to be the first to go out. But it does not necessarily mean that the exact oldest physical object has been tracked and sent out. In other words, the cost associated with the inventory that was purchased first corresponds to the first cost expenses. With FIFO management, the cost of the inventory represents the cost of the materials most recently purchased.

Complimentary to the preceding technique is last-in first-out (LIFO), which means that the most recently purchased items are recorded to go out first.

The behavior of a queue depends on each specific situation and can be associated with a specific stochastic distribution that best fits it, as will be synthesized next.

The exponential distribution is one of the major tools in queuing theory. The exponential random variables possess the memoryless property, which makes the analysis of such models easily manageable. This distribution is associated with the Poisson process.

In fact, the occurrence of sequenced discrete events can often be realistically modeled by a Poisson process. The defining characteristic of such a process is that the time intervals between successive events are exponentially distributed. Given a sequence of discrete events occurring at times $t_0, t_1, t_2, t_3, ..., t_n$, the intervals between successive events are the following:

$$\Delta t_1 = (t_1 - t_0), \Delta t_2 = (t_2 - t_1), \Delta t_3 = (t_3 - t_2), ... , \tag{11.7}$$

In a Poisson process, these intervals are managed as independent random variables drawn from an exponentially distributed population, that is, a population with the density function

$$f(x) = \lambda e^{-\lambda x} \tag{11.8}$$

for some fixed constant λ.

The exponential distribution is particularly convenient for mathematical modeling, because it implies a fixed rate of occurrence.

Several references can be consulted about this subject, such as the following: Haviv (2013) and Robert (2013).

11.6.7 Ant Algorithm

The ant algorithm, or the ant colony algorithm, is a population-based approach for solving combinatorial optimization problems that are inspired by the searching behavior of ants and their inherent ability to find the shortest path from a food source to their nest.

The fundamental approach underlying the ant algorithm is an iterative process in which a population of simple agents repeatedly construct candidate solutions. This construction process is probabilistically guided by heuristic information about the problem under analysis as well as by a shared *memory* containing the experience gathered by the ants in the previous iteration.

In the ant colony optimization algorithm, the problems are defined in terms of *components* and *states* (sequences of components). The algorithm incrementally generates solutions in the form of paths in the space of components, adding new components to a state. The memory is kept for all the observed transitions between the pairs of solution components. It is associated with a degree of desirability for each transition, depending on the quality of the solutions in which it has happened so far. While a new solution is generated, a component α is included in a state, with a probability that is proportional to the desirability of the transition between the last component included in the state and α itself. From that point of view, all states finishing with the same component are identical.

The typical situation where the ant algorithm can be used is in the traveling salesman problem, and it can be extrapolated to maintenance planning both in internal logistics and with dispersed facilities, for example, when a number of activities have to be processed on the same equipment and can only be processed one activity at a time. Before an activity can be processed, the equipment must be prepared. Given the processing time of each activity and the switch-over time between each pair of activities, the objective is to find an execution sequence of the activities that makes the total intervention time as short as possible.

Several references can be consulted about this subject, such as the following: Bonabeau et al. (1999) and Dorigo and Stützle (2004).

11.6.8 Dijkstra's Algorithm

Dijkstra's algorithm permits finding the shortest path between nodes in a graph. The problem of finding the best path between two points is based on graph theory, where there are several algorithms. One of them, Dijkstra's algorithm, is a good approach when the costs associated with the trajectories are being evaluated and they are positive. In the case where the costs associated with the paths exhibit negative values, Dijkstra's algorithm does not work, with one possible solution through graph theory given by Bellman-Ford's algorithm.

The synthesis for the approach of Dijkstra's algorithm that tries to find the solution to the single-source shortest path problem, based on graph theory, is the following:

- Both directed and undirected graphs.
- All edges must have nonnegative weights.
- Graphs must be connected.

Bellman-Ford's algorithm, as mentioned before, has the following characteristics:

- Works for negative weights
- Detects a negative cycle if any exists
- Finds the shortest simple path if no negative cycle exists

11.7 A Case Study

The case study presented here corresponds to a problem of a maintenance logistics, with the objective of planning and assigning wind farm maintenance teams, which uses genetic algorithms, namely those discussed in the last two subsections. Only some highlights will be given here, because the case study is published in a paper where the author participated as supervisor (Fonseca et al., 2014).

The optimization problem of the routes between the different wind farms is important because of the costs of the trips and the resources involved. Based on the knowledge of the wind farm locations, the intent is to optimize the distances covered, having as conditioning factors the following: costs, namely losses of energy production; estimated wind speed for a given time; availability of high-cost equipment required for interventions; intervention times; travel times; and so on. There is also another restriction, that is, the number of alternative routes, which is limited among the different parks when they are installed in mountainous areas with difficult access. The opposite is true for the sea parks where, by air, the journey may be in a straight line or, in the case of a sea route, may also depend on the navigability of the routes and the routes themselves.

Considering a GPS system and the terrestrial geodetic model with GPS points, it is possible to estimate the distances and, even with noncataloged geographic locations, the points of the road intersections to build the best wind farm access can be determined.

Each one of the points can be grouped in two distinct groups:

$$G = \{(Lat_1, Longt_1), \dots , (Lat_n, Longt_n)\} \tag{11.9}$$

with

$$(Lat_i, Longt_i) \in \{Wind\ generators\ geographical\ position\}$$

and the set

$$C = \{(Lat_1, Longt_1), \dots , (Lat_n, Longt_n)\} \tag{11.10}$$

with

$$(Lat_i, Longt_i) \in \{Geographical\ position\ between\ wind\ generators\}$$

where the set C (routes) aggregates the constraints imposed on the elements of the set G (wind generators) to travel between any two elements.

Figure 11.1 graphically illustrates the problem. The use of two maintenance teams is assumed. The goal is to determine what the sequence of maintenance interventions of the wind generators should be.

To solve the problem, several visits to the same node may be necessary if the number of the maintenance intervention days is over one. The problem presented in Figure 11.1 may be decomposed into two basic problems:

- To choose the best path between two nodes
- To choose the sequence of visits to generators with faults (red) or predicted failures (light and dark yellow), depending on the wind forecast and the sale price of energy, among others

The cost function is expressed in monetary units representing the financial cost associated with the loss of electricity production and the cost of human resources and logistics, such as travel and overnight stays. The objective is to maximize the balance between credits and debits.

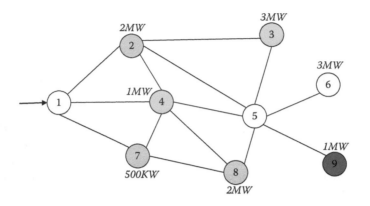

FIGURE 11.1
Wind generator locations with the following situations: red—fault; light and dark yellow—probability of future failure with different priorities.

The following parameters must also be considered:

- Maximum power available in a wind generator
- Number of hours of production per day
- Price per each MWh produced
- Number of estimated days for a failure to occur

TABLE 11.1

Solutions about the Best Routes between the Origin and Destination Node

Cost	Genetic	Dijkstra	Ant	Origin	Destination	1st	2nd	3rd	4th
80.3	0.1101584	0.0000001	0.7310512	1	9	1	2	5	9
80.3	0.1101584	0.0000001	0.4506480	9	1	9	5	2	1
20.1	0.1001440	0.0000001	0.6309072	5	9	5	9	–	–
20.1	0.1101584	0.0000001	0.1301872	9	5	9	5	–	–
90.3	0.1402016	0.0000001	0.6609504	1	6	1	2	5	6
90.3	0.1101584	0.0000001	.4506480	6	1	6	5	2	1
30.1	0.1101584	0.0000001	0.5507920	5	6	5	6	–	–
30.1	0.1101584	0.0000001	0.1301872	6	5	6	5	–	–
20.1	0.1101584	0.0000001	0.2002880	1	2	1	2	–	–
20.1	0.1101584	0.0000001	0.2503600	2	1	2	1	–	–
40.1	0.1101584	0.0000001	0.3004320	5	2	5	2	–	–
40.1	0.1101584	0.0000001	0.3605184	2	5	2	5	–	–
60.1	0.1101584	0.0000001	0.4005760	1	7	1	7	–	–
60.1	0.1101584	0.0000001	0.3905616	7	1	7	1	–	–
70.2	0.1101584	0.0000001	0.4806912	5	7	5	4	7	–
70.2	0.1001440	0.0000001	0.3505040	7	5	7	4	5	–
110.2	0.1101584	0.0000001	0.4806912	1	8	1	7	8	–
110.2	0.1101584	0.0000001	0.5207488	8	1	8	7	1	–
60.1	0.1101584	0.0000001	0.5708208	5	8	5	8	–	–
60.1	0.1101584	0.0000001	0.3304752	8	5	8	5	–	–
60.2	0.1101584	0.0000001	0.3605184	5	1	5	2	1	–

- Number of maintenance intervention days for the estimated failure
- Loss factor when a wind generator is under maintenance
- Number of teams available simultaneously to carry out the work

It is not considered possible for two or more teams to work simultaneously on the same aero generator.
Each team has autonomous transportation:

- Nodes where there are overnight stays
- Price per night per team at a certain node

During the journeys, it is assumed that each team has an autonomous means of transport. To estimate the travel costs, the following parameters are taken into account:

- Price per kilometer of travel in each route
- Number of kilometers of each route
- Number of travel days between two adjacent nodes
- Cost paid to each team for travel between two adjacent nodes of a route

The solutions searched for the problem under discussion were found from a genetic algorithm, Dijkstra's algorithm, and the ant algorithm. The results are shown in Table 11.1.
All the algorithms find the optimal solution, but with different computational costs. Dijkstra's algorithm is the most efficient, followed by the genetic algorithm and the ant algorithm. However, the latter is able to better adapt to a temporally dynamic cost matrix.

12

Condition Monitoring

12.1 Background

Maintenance by condition monitoring (CM), which may also be predictive, is a method of maintenance planning that can be applied to any physical asset, whether of industry, aeronautics, oil refineries, ships, power stations, or hospitals, among others. CM is supported by a set of technologies, namely those developed in the last decades, such as those related to the use of sensors and monitoring of the asset performance. CM allows very significant gains in the asset's availability, performance, and, as a consequence, productivity.

This type of maintenance, when compared to systematic maintenance, presents the need to reinforce the diagnostic capacity in order to be able to closely follow the state of the "health" of the assets.

The current techniques of CM involve the use of several knowledge tools, such as vibrations, lubricants, electrical parameters, and effluent analysis, among others. However, it should be noted that there are specific techniques that are very relevant to the monitoring of equipment conditions based on the control variables of the production process, as is the case of those based on Six Sigma control charts.

In general, none of the condition control techniques can be considered enough. The complementary analysis that results from following several variables of each asset gives the final prognosis that permits an increase of asset functioning time until the next maintenance intervention.

In fact, a particular technique may cover a wide range of potential causes of damage but, in general, is not enough to ensure the detection of all possible causes of degradation. Therefore, there is usually complementarity among several techniques.

12.2 Techniques for Condition Monitoring

The next sections present some of the most common techniques for condition monitoring. The objective is to describe some of the most-used tools and demonstrate their contribution to the reliability and availability of the assets. The first one to be presented is vibration analysis.

12.2.1 Vibration Analysis

A body is vibrating when it describes an oscillating movement around a reference position. The number of complete movements (cycles) per second is called the frequency, measured in Hertz (Hz = cycles/second).

Mechanical vibrations can be generated to produce useful work, such as in vibratory feeders, impact crushers, compactors, concrete vibrators, and many movements necessary to the industry sector.

However, vibrations are often undesirable, which indicates that their presence may cause unplanned downtime. An increased vibration level is related to changes in one or more elements of the equipment, which may influence other interconnected components.

A small vibration can excite resonant frequencies in other structural parts and be amplified to a higher level of vibration, which can be felt in the structure and not directly at the source of vibration.

Vibration monitoring is a knowledge area with much research, development, and applications around the world. At the end of this section, some references will be supplied for a deeper study of this subject.

To conduct vibration analysis, some aspects must be considered, like the following:

- The parameters to be evaluated—Vibration parameters to be collected at each point, initial (signature) and alarm levels, measurement points, and routes.
- Choice of equipment to be analyzed—Class A equipment, vital for the production process, is the first to be analyzed; the incorporation of equipment of other classes, namely B, may be justified.
- Measurement points—The vibrations originating from rotating parts may be felt in the static parts and are transmitted from the bearings.

The are many causes of undesirable vibrations in equipment, some of which are the following:

- Imbalance of masses
- Misalignment of axes
- Generalized slack
- Gear teeth damage
- Bearings damage

The vibration signals of equipment give information related to its operation, indicating its health and giving data to support the decision about possible maintenance intervention in it.

Each equipment has a characteristic shape of vibration in aspect and level (vibration signature). However, equipment of the same type may exhibit variations in its dynamic behavior. This is due to variations in settings, tolerances, and, mainly, defects.

Each element of an equipment induces its own excitation, generating a specific perturbation. These elements are, for example, rotors, gears, bearings, and so on.

The dynamic behavior of equipment is a composition of the contribution of all its components, including defects and excitation from the moving parts.

Vibrations occur at various frequencies due to various excitations. The movement at each point is the overlapping of several harmonics.

For equipment maintenance purposes, the diagnosis to identify the causes of the anomalies is obtained by separating the harmonics from the global signal in relation to the equipment vibration signature and associating them with the defective elements.

Some of the main effects of vibrations are the following:

- High risk of accidents
- Premature wear of components
- Unexpected breaks
- Increased maintenance costs
- Structural fatigue
- Decreased availability

When attempting to identify faults in rotating motors, a change in the vibration signal can be considered a change in the equipment condition. Vibrations tend to change with machine speed and load.

Some of the most important variables in vibration condition monitoring are the following:

- Peak-to-peak value (V_{pp})—Corresponds to the difference in the module between the maximum and minimum values when the positive and negative peaks are symmetric
- Peak value—The maximum value in both the positive and negative peak:

$$V_p = \frac{V_{pp}}{2}$$

- Effective value (root mean square—RMS)—The value that corresponds to the energy spent to perform work

$$V_{RMS} = \frac{1}{T} \int_0^T f^2(t) \cdot dt$$

- Average value—Corresponds to the arithmetic mean overall values in a cycle

$$V_{avg} = \frac{1}{T} \int_0^T f(t) \cdot dt$$

Any periodic waveform, that is, for which $f(t) = f(t + T)$, can be expressed by a Fourier series, as long as it obeys the Dirichlet conditions, which are the following:

- If the waveform is discontinuous, it has a finite number of discontinuities in the period T.
- The waveform has a finite mean value in the period T.
- The waveform has a finite number of positive and negative maxima.

Once the preceding conditions are satisfied, the Fourier series is given by:

$$f(x) = \frac{a_0}{2} + \sum_{n=1}^{\infty}\left(a_n \cos\frac{n\pi x}{L} + b_n \sin\frac{n\pi x}{L}\right)$$

The Fourier coefficients a_n and b_n are given by the following expressions:

$$a_n = \frac{1}{L}\int_{-L}^{L} f(x)\cos\frac{n\pi x}{L}dx, \quad (n = 0,1,2,3...)$$

$$b_n = \frac{1}{L}\int_{-L}^{L} f(x)\sin\frac{n\pi x}{L}dx, \quad (n = 1,2,3...)$$

The function $\cos(x)$ is even, while the $\sin(x)$ function is odd.

If $f(x)$ is even, then $(f(-x) = +f(x)$ for every $x)$, which implies that $b_n = 0$ for all n, leaving only the Fourier series with the cos function (and perhaps a constant term) to the function $f(x)$.

If $f(x)$ is odd, then $(f(-x) = -f(x)$ for every $x)$, then $a_n = 0$ for all n, leaving the Fourier series only with the function sin to the function $f(x)$.

The are many standards concerning vibration analysis, namely the following:

- IEC 60034-14, *Mechanical vibration of certain machines with shaft heights 56 mm and higher—Measurement, evaluation, and limits of vibration severity*
- NEMA MG 1—*Part 7, Mechanical vibration measurement, evaluation and limits*
- IEEE 841 standard for petroleum and chemical industry—*Severe duty totally enclosed fan-cooled (TEFC) squirrel cage induction motors—Up to and including 370 kW (500 hp)*
- BS 4999-142, *General requirements for rotating electrical machines, specification for mechanical performance: Vibration*
- API 541, *Form-wound squirrel-cage induction motors—500 hp and larger*
- API 546, *Brushless synchronous machines—500 kVA and larger*
- API 547, *General-purpose form-wound squirrel cage induction motors—250 hp and larger*
- API 670, *Machinery protection systems*
- API 684, *Rotor dynamics tutorial: Lateral critical speeds, unbalance response, stability, train-torsional and rotor balancing*
- ISO 1940-1, *Mechanical vibration—Balance quality requirements for rotors in a constant (rigid) state—Part I: specification and verification of balance tolerances*
- ISO 1940-2, *Mechanical vibration—Balance quality requirements of rigid rotors—Part 2: Balance errors*
- ISO 2954, *Mechanical vibration of rotating and reciprocating machinery—requirements for instruments for measuring vibration severity*
- ISO 7919-1, *Mechanical vibration of non-reciprocating machines—measurements on rotating shafts and evaluation criteria—Part 1: General Guidelines*

- ISO 7919-3, *Mechanical vibration of non-reciprocating machines—measurements on rotating shafts and evaluation criteria—Part 3: Coupled industrial machines*
- ISO 7919-5, *Mechanical vibration of non-reciprocating machines—measurements on rotating shafts and evaluation criteria—Part 5: Machines sets in hydraulic power generating and pumping plants*
- ISO 8528-9, *Reciprocating internal combustion engine driven alternating current generating sets—Part 9: Measurement and evaluation of mechanical vibrations*
- ISO 8821, *Mechanical vibration—Balancing—Shaft and fitment key convention*
- ISO 10814, *Mechanical vibration—Susceptibility and sensitivity of machines to unbalance*
- ISO 10816-1, *Mechanical vibration—Evaluation of machine vibration by measurements of non-rotating parts—Part 1: General Guidelines* (replaces the old VDI/ISO 2372)
- ISO 10816-3, *Mechanical vibration—Evaluation of machine vibration by measurements of non-rotating parts—Part 3: Industrial machines with nominal power above 15 kW and nominal speeds between 120 and 15.000 rpm when measured in situ*
- ISO 10817-1, *Rotating shaft vibration measuring systems—Part 1: Relative and absolute sensing of radial vibration*
- ISO 13373-1, *Condition monitoring and diagnostics of machines—vibration condition monitoring—Part 1: General procedures*
- ISO 13373-2, *Condition monitoring and diagnostics of machines—vibration condition monitoring—Part 2: Processing, analysis and presentation of vibration data*
- ISO 15242-2, *Rolling bearings—Measuring methods for vibration—Part 2: Radial ball bearings with cylindrical bore and outside surface*
- ISO 15242-3, *Rolling bearings—Measuring methods for vibration—Part 2: Radial spherical and tapered roller bearings with cylindrical bore and outside surface*
- ISO 16063-1, *Methods for the calibration of vibration and shock transducers—Part 1: Basic concepts*
- ISO 16063-21, *Methods for the calibration of vibration and shock transducers—Part 21: Vibration calibration by comparison with a reference transducer*
- ISO 18436-2, *Condition monitoring and diagnostics of machines—requirements for training and certification of personnel—Part 2: Vibration condition monitoring and diagnostics*
- ISO 20806, *Mechanical vibration—Criteria and safeguards for the in situ balancing of medium and large rotors*

There are many important references in this area, such as Goldman (1999), Scheffer and Girdhar (2004), Adams (2009), Sinha (2014), and Kelly (2006).

12.2.2 Oil Analysis

Lubricating oils are substances placed between two movable surfaces or between a fixed and a movable one. They form a protective film whose main function is to reduce friction and wear, as well as to assist in controlling the temperature and sealing the components of machines and motors. This provides cleaning of the parts, protects against corrosion due to oxidation processes, and prevents the entrance of impurities. They can also be agents of transmission of force and movement.

The main functions of a lubricating oil are the following:

- Limit friction (grease)—Reduce or eliminate friction.
- Reduce premature wear and power consumption—Reduce or eliminate premature wear. In special situations, provide controlled wear (bearing housing, for example). Keep power consumption as low as possible.
- Limit the temperature (motors)—Cooling effect.
- Limit corrosion—Usually avoid corrosion.
- Isolate electrical currents—In electrical equipment, but also in motors and other equipment.
- Transmit force—In hydraulic systems.
- Damping shocks—In shock absorbers and gears.
- Remove contaminants—Wash motors and keep them clean.
- Seal (motors, hydraulic systems)—In motors, avoid the escape of gases from the combustion chamber. In other equipment, for example, avoid the input of impurities.
- Transport waste (cutting fluids, motors)—In metal cutting waste, from internal equipment wear, for example.

Lubricating oils may have two main failure processes:

- The first occurs due to the contamination by particles of wear of the equipment or by external agents, with water being one of the most common contaminants in industrial assets.
- The second is related to the degradation of properties due to changes in the characteristics of the lubricant, damaging the performance of its functions.

There are a lot of rules and entities that must be taken into account in the lubricating oils field, namely the following:

- The American Society of Automobile Engineers (SAE):
  - The SAE has a grade that classifies oil's fluidity at high and low temperatures. Its purpose is to classify lubricating oils for engines and transmissions on the basis of their viscosity at a reference temperature.
- The European Automobile Manufacturers' Association (ACEA):
  - Some of the ACEA's members are the following:
    - Ford Europe, PSA Peugeot-Citroën, Daimler, BMW, Jaguar-Land Rover, Renault Group, DAF Trucks, Fiat, GM Europe, MAN, Porsche, Scania, Volkswagen, Volvo and Toyota Europe
  - The ACEA has created a classification of lubricants according to its technical specifications and the requirements of each engine type. Tests are performed in order to classify lubricants into standardized categories, primarily using European engines and under European driving conditions. The ACEA standard is made up of a letter that represents the engine type and a number that represents its performance. The latest version of the ACEA standard defines:

- – Four categories of common standards for gasoline (letter A) and diesel (letter B) engines
- – Four categories for vehicles equipped with post-treatment systems (letter C)
- The American Petroleum Institute (API):
  - The API is the American organization that represents the petroleum and natural gas industry. It is made up of petroleum companies, gasoline additive firms, car manufacturers, and testing laboratories. Its role is to create a classification according to the performance of lubricants.
  - The API standard uses two letters:
    - – The first represents the type of application (S is for service-qualifying gasoline engines, C for commercial-qualifying diesel engines).
    - – The second gives the lubricant performance level, according to the year the standard entered into effect.
- The International Lubricant Standardization and Approval Committee (ILSAC):
  - The ILSAC is the organization responsible for creating lubricant specifications for passenger cars. ILSAC/OIL is the entity within ILSAC that develops and introduces newly required specifications. The entity is divided into two branches: ILSAC, which includes the Alliance of Automobile Manufacturers (AAM) and the Japanese Automobile Manufacturers Association (JAMA).
- The Japanese Automotive Standards Organization (JASO):
  - The JASO has established its own standards in terms of performance and quality for Japanese engines.
  - The JASO classifies oils into three categories:
    - – DH-1
    - – DH-2, for industrial diesel engines
    - – DL-1, for diesel passenger cars engines and qualification of fuel economy lubricants
- The Euro standards:
  - European Community standards specify the maximum limits for heavy-duty vehicle pollution emissions:
    - – For passenger cars, there are two Euro standards in six categories: Euro 6B (2014) and Euro 6C (2017).
  - To constantly improve air quality, the Euro standards take several factors into account when evaluating a vehicle's pollution (the level of carbon monoxide, the different nitrogen oxides, fine particle emissions, etc.),
  - The Euro standards must be directly applied by the automotive manufacturers.

12.2.3 Other Techniques

There are many other techniques beyond vibration and oil analysis that can be applied to monitor equipment's health state, like the following:

- Temperature
- Effluents:

- CO_2, CO
- Volatile organic compounds (VOCs) or total of hydrocarbons (THC)
- NO_x, particularly NO_2
- SO_2, NH_3, N_2O, PM_{10}, $PM_{2.5}$, Pb
- Other heavy metals like Cd, Zn, Cu, Cr, Ni, and Se
- H_2S
- Electrical currents, voltage, and power
- Structural health monitoring (SHM)

The last technique referred to, structural health monitoring, can be emphasized because of its increasing importance. SHM is the process of implementing a damage detection and characterization strategy for engineering structures.

Damage is defined as changes to the material and or geometric properties of a structural system, including changes in the boundary conditions and system connectivity, that adversely affect the system's performance.

SHM applications are increasing, being accelerated by aeronautics and eolic towers, because it permits continuous, autonomous, in-service monitoring of the physical condition of a structure through embedded or attached sensors with a minimum of manual intervention to monitor the structure's structural integrity.

SHM is implemented based on fixing permanent sensors on the structure. SHM includes all monitoring aspects related to damage, loads, conditions, and so on. The sources of faults result from fatigue, corrosion, impacts, excessive loads, unforeseen conditions, and so on.

Some features of the SHM are the following:

- The sensors are permanently applied to the structure.
- Physical access to the inspection area is not necessary.
- Manual operation in the inspection area is not required.
- There is safe inspection in hazardous areas.
- The use of scanners is not necessary.
- There is an automated inspection.
- Several locations may be analyzed at the same time.
- There is no influence of the human factor on inspection results.
- The final objective is to imitate the human nervous system.

The monitoring process results in a lot of data that can be processed several ways for diagnostic and prognostic purposes. As referred to in this book, several tools can be used, like time series analysis, artificial neural networks (ANNs), and finite element analysis (FEA), with the objective of generating the simulated situations required for NN training.

12.3 Types of Sensors

There are many types of sensors, according to each type of condition variable. The reasons to choose a sensor are diverse, as are the type of output signal, the range of output values, the environmental conditions, the physical dimensions, and so on.

Sensor signals may be digital or analog. If they are digital, they have an interface to communicate with the other devices. If they are analog, an analog-to-digital converter is needed to make it possible for them to communicate with other devices.

There are several manufacturers of sensors, as was discussed in Chapter 9, namely in Table 9.3.

Bogue (2013) presents a review of technologies and applications of sensors for condition monitoring.

12.4 Data Acquisition

Data acquisition of condition variables can be done in several ways. The classical approach is through readings taken by technicians, usually periodically, with the aid of special equipment and or by taking samples. Classical examples of this are thermography, noise, vibrations, and so on.

The current tendency is the use of sensors connected online using specific or commercial networks.

In both situations, the data must be stored and evaluated. It is through their tracking and analysis, and with the critical reference levels well defined, that maintenance interventions can be done before the variables reach values that may correspond a fault.

If the degradation of the values of condition monitoring variables is analyzed by specific algorithms to predict the next values, the next maintenance intervention can be forecast in advance, usually permitting maximization of the interval between interventions—this methodology is called predictive maintenance.

12.5 On-Condition Online

For data acquisition online, sensors like those referred to in Section 12.3 and Chapter 9 can be used. Through some additional electronics, they are able to communicate and, at the end, a CMMS can report a working order before the next fault.

The tendency is to have, time after time, smaller, intelligent sensors with ability to communicate via both wire and/or wireless with the CMMS and other tools—these are IoT sensors.

There are a lot of sensors able to communicate autonomously, namely the ones referred to in the ambit of condition monitoring. Additionally, a lot of other types of sensors that go into IoT devices can be mentioned:

- Chemical/gas sensors
- Gas identification sensors
- Force/load/torque/strain sensors
- Heat sensors
- Humidity/moisture sensors
- Motion/velocity/displacement/position sensors
- Presence/proximity sensors

- Pressure sensors, transducers
- Temperature sensors
- Tilt switches
- Vibration and shock sensors
- Water quality

An IoT network of devices connects them directly to each other to capture and share condition data through a secure service layer (SSL) connected to a server, usually in the cloud. It combines sensors, microcontrollers, microprocessors, and gateways where sensor data are further analyzed and sent to the cloud, then to technicians or other users. This subject will be discussed more deeply in Section 12.7.

12.6 On-Condition with Delay

One problem that must be taken into account is the occupation of commercial networks when using remote Wi-Fi sensor measurement. Additionally, in many situations, there are no reasons to be continuously sending and receiving data from sensors. For example, the following situations may occur:

- The variation of values of signal sensors is too slow; then there is no need to continuously send data to the server.
- The communications cost is too high and the equipment condition is compatible with a low sample reading.
- The communications are not stable and it is necessary to store data in the firmware of the sensor and send the data when communications happen.

In the preceding situations and many others, it would be necessary to manage the data according to the real situation and the specificity of the asset, what means that even when the systems have connections to read the values of sensors online, in practice, these data are read with a delay.

12.7 Technology for Online Condition Monitoring

For online condition monitoring, it is necessary to have a network to connect the sensors and respective systems to the server.

The first networking technology for an IoT device is Wi-Fi. However, Wi-Fi needs a fair amount of power, and there are a myriad of devices that cannot afford that level of power, for example, sensors placed in locations that are difficult to power from the grid. Thus, it is important to consider low-power solutions.

Newer networking technologies allow the development of low-cost, low-power solutions, supporting the creation of very large networks of very small intelligent devices. Currently, major R&D efforts include:

- Low-power and efficient radios, allowing several years of battery life
- Energy harvesting as a power source for IoT devices
- Mesh networking for unattended long-term operation without human intervention (e.g., machine-to-machine [M2M] networks)
- New application protocols and data formats that enable autonomous operation

One of the major IoT enablers is the IEEE 802.15.4 radio standard, released in 2003. Commercial radios meeting this standard provide the basis for low-power systems. This IEEE standard was extended and improved in 2006 and 2011 with the 15.4e and 15.4g amendments. Power consumption of commercial RF devices is now cut in half compared to only a few years ago, and another 50% reduction is expected with the next generation of devices.

IPv6 over low-power wireless personal area networks (6LoWPAN) is another possible solution because the devices that take advantage of energy harvesting must perform their tasks in the shortest time possible, which means that their transmitted messages must be as small as possible. This requirement has implications for protocol design. This is one of the reasons 6LoWPAN has been adopted by ARM (Sensinode) and Cisco (ArchRock). 6LoWPAN provides encapsulation and header compression mechanisms that allow lower transmission times.

Regarding some wireless radio technologies, Section 9.4 of Chapter 9 covered some of this subject.

12.8 Technology for Offline Condition Monitoring

The most traditional way to measure condition variables on equipment to monitor the condition state is through systematic readings made locally, close to the equipment, as is the case of vibration analysis, temperature, effluents, and so on.

When condition variables are read by the technicians, it is important to pay special attention to some aspects like the following:

- To make systematic measures at periodic time intervals
- To make measures at the same points for the same equipment

Additionally, some variables involve extra operations like oil analysis because it includes taking oil samples to specific recipients to be analyzed with specialized equipment.

Most equipment permits making a first diagnosis immediately after the measurement made by the test and measurement equipment. However, both with this and/or through a computer, it is usually possible to perform a deeper analysis through the historical data and, aided by this and/or additional math, to forecast the next value of the condition variable.

The usual solutions for offline condition monitoring, as noted before, are many. Next, only some of the most useful will be discussed.

The first one is vibration analysis, which is done through sensors, usually accelerometers, that are placed on specific points of the equipment, like bearings or other points where the vibrating signal is more easily and accurately measured. These sensors are usually

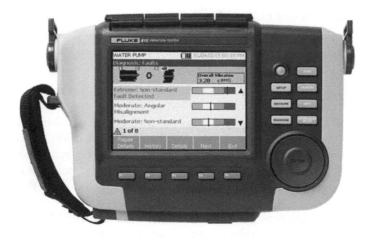

FIGURE 12.1
Fluke 810 vibration tester. (From http://en-us.fluke.com/products/vibration-meters-and-laser-alignment-tools/fluke-810-vibration.html.)

FIGURE 12.2
GE portable vibration analyzers. (From https://www.gemeasurement.com/condition-monitoring-and-protection/portables/bently-nevada-scout-vbseries-portable-vibration.)

connected to equipment that registers and stores the data, allowing comparison of the present measures with the historical values.

Additionally, equipment for tests and measurements allows transmitting data to a computer through the universal serial bus (USB) port, Bluetooth, or others. Figures 12.1 and 12.2 show some equipment for vibration monitoring.

Another common technique for condition monitoring is through the temperature variable, both the temperature itself and also the thermographic image. The next figures (Figures 12.3 and 12.4) show some commercial equipment.

FIGURE 12.3
Thermographic camera Flir TG165. (From http://www.flir.com/instruments/tg165/.)

FIGURE 12.4
Fluke TiS10 infrared camera. (From http://www.fluke.com/fluke/uken/thermal-imaging/Fluke-TiS10.htm?PID=79858.)

Oil analysis is one of the most important techniques for condition monitoring. Instead of measuring the oil variables *in situ*, oil samples are usually collected that are later analyzed using special equipment. Figure 12.5 shows an example of a receptacle to collect oil samples.

There are many other techniques that may be used offline through specific equipment, like the above mentioned. The next reference allows deeper exploration of this subject: Mohanty (2014).

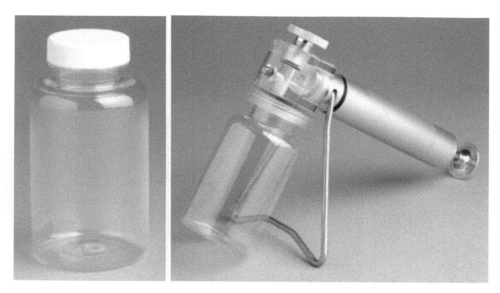

FIGURE 12.5
Oil sampling bottles. (From https://www.tricocorp.com/product/oil-sampling-bottles/.)

12.9 Augmented Reality to Aid Condition Maintenance

Augmented reality (AR) is a technology that was born in the industrial environment with the objective of providing digital intuitive instructions at the same time as technicians were working on their tasks in order to reduce time spent looking for instructions in equipment manuals (Sims, 1994).

AR is a technology that superimposes virtual data over a real environment. There are a lot of variations of real AR systems, including the configuration of commercial devices, in which the virtual contents can be displayed through 2D data or 3D models. Human–machine interaction (HMI) can be more or less friendly, according to the way the target locations are identified on the scene.

AR solutions may use two types of approaches: the classical one that uses markers and the current tendency that tries to be markerless.

The use of markers represents a huge restriction for AR in industrial environments, because the dust and pieces position themselves. Thus, objects should be found based on their natural characteristics, including shape, texture, or edges.

Additionally, for daily use, the AR system's hardware ought to be portable, because one of the AR objectives is to process all the data on tablets or smartphones, because they are small enough to be carried naturally by technicians (Oliveira et al., 2014).

After recognizing the equipment, module, or component, the AR system can search the database that contains the historical values of condition variables and monitor the equipment's health and/or process prediction algorithms in order to immediately show what is expected for the future of the equipment condition. If necessary, the system may launch a working order.

12.10 Holography

Holography is a technique still in primary development. However, it is a very promising tool for the near future. In fact, it adds a powerful potential for technicians to go around the equipment, module, or piece and eventually interact with it.

Acoustic holography has been used since several years ago, as can be seen in several case studies. The case presented by Takahashi et al. (1999) is only one example.

Regarding color holography, there are some preliminary experiences, as described by Ikegami et al. (2001) in an application in nuclear power plants.

12.11 A Case Study

At this point, a highlight of oil condition monitoring of public transport diesel engines is presented in three steps:

1. First, the monitoring was done through periodic oil samples collected to be analyzed offline.
2. Second, the results obtained from the analysis and the algorithms used were studied.
3. Third, an analysis of condition monitoring maintenance was done.

In this context, the monitoring of the evolution of the oils' degradation was made for three homogeneous bus groups.

The fuel of these vehicles is composed of a mixture of 30% of biodiesel. One objective is to assess the effect of the mixture on the degradation of the lubricants as well as on the changes in the maintenance of these vehicles, which use two types of oils, but with similar characteristics of temperature:

- Lubricant A—10 W 40
- Lubricant B—10 W 40

The features and conditions of functioning of the lubricants used (provider/brand) are synthesized in the next points:

Lubricant A:

- The oil is synthetic, multigrade extra-high-performance diesel oil (EHPDO), especially recommended for diesel engine lubrication of high-power heavy vehicles with natural aspiration or turbocharged, operating in the most severe conditions, including very large changing oil gaps, having the following properties (Table 12.1):
  - High stability of the lubricant film and maintenance of its properties, even under adverse pressure and temperature conditions
  - Detergent/dispersant capacity reinforced, ensuring perfect cleaning of the engine by inhibiting the formation of deposits in the segment boxes, lacquers, and the skirts of the pistons

TABLE 12.1

Main Features of Lubricant A

Properties	
SAE grade	10 W 40
Density at 15°C, Kg/I (D1298/D4052)	0.872
Viscosity index (D2270)	139
Kinematic viscosity at 40°C, mm²/s (D445)	107.2
Kinematic viscosity at 100°C, mm²/s (D445)	14.5
Inflammation point, °C (293)	197
Fluxion point, °C (D97/D6892), Max.	−39
Basicity number, mg KOH/g (D2896)	12.5

- Excellent antiwearing properties
- Good pour point, ensuring that at start-up, the oil quickly reaches all components to be lubricated
- High antioxidant point
- Alkaline reserve with high stability throughout the oil life

Lubricant B:

- The oil is a multigrade lubricant with synthetic basis, type ultra-high-performance diesel oil (UHPDO), specially designed for the lubrication of diesel engines of heavy vehicles of high power, atmospherics, or turbocharged operating in extreme conditions. The use of this kind of lubricants allows a considerable reduction in the fuel consumption and achieves large intervals of oil replacement. This product follows the new ACEA E7-04 specifications, with the properties presented in Table 12.2:
  - High stability of the lubricant film and maintenance of its properties, even under adverse pressure and temperature conditions
  - Improved detergent/dispersant capacity, ensuring perfect cleaning of the engine through the inhibition of deposit formation
  - Alkaline reserve with high stability throughout the oil life
  - Good flowability at low temperatures, facilitating cold start

Table 12.3 shows an example of a data analysis from an oil motor bus.

TABLE 12.2

Main Properties of Lubricant B

Properties	
SAE grade	10 W 40
Density at 15°C, Kg/I (D1298/D4052)	0.87
Viscosity index (D2270)	141
Kinematic viscosity at 40°C, mm²/s (D445)	100
kinematic viscosity at 100°C, mm²/s (D445)	13.9
Inflammation point, °C (293)	197
Flow point, °C (D97/D6892), Max.	−36

TABLE 12.3

Lubricant Sheet with the Analysis of an Oil Bus

Lubricant Analysis					
Fleet number	282				
Equipment Data					
Registration	00-00-XX		Brand	AAA	Model
Lubricant Data					
					√—Normal
Lubricant	EC 10 W 40				Δ—Alert
					X—Danger
Sample Results					
Date			17-07-07	26-11-07	10-03-08	25-03-09
Sample reference			370,391	391,740	408,637	470,969
Kms of equipment			184,438	197,707	209,796	258,683
Kms of lubricant			9832	10,000	10,000	15,000
Condition						
Antifreeze (%)	(PE-TA.071)	0.08	0.08	0.08	0.08
Appearance	(PE-TA.096)	Black	Black	Black	Black
Fuel (%)	(PE-TA.071)	2	2	2	2
Water content (%)	(PE-TA.071)	0.1	0.14	0.1	0.1
Water content (FinachecK) (%)	(PE-5022-Al)				
Soot (%)	(DIN 51452)	2.7	3.1	3	0.8
Nitration (ABS/cm)	(PE-TA.071)	6	5	4	1
Oxidation (ABS/cm)	(PE-TA.071)	1	1	4	1
Sulfidation (ABS/cm)	(PE-TA.071)	5	6	1	1
TBN (mgr KOH/gr)	(ASTM D-2896-07a)	10.83	10.3	10.55	10.2
Viscosity at 100°C (cst)	(ASTM D-445-11)	13.9	14	13.7	12.8
Wear and Contamination Metals						
Content in Al (ppm)	(ASTM D-5185-05 mod.)	2	2	1	3
Content in Cr (ppm)	(ASTM D-5185-05 mod.)	1	1	1	0
Content in Cu (ppm)	(ASTM D-5185-05 mod.)	1	2	1	4
Content in Fe (ppm)	(ASTM D-5185-05 mod.)	22	22	19	12
Content in Mo (ppm)	(ASTM D-5185-05 mod.)	2	3	2	4
Content in Na (ppm)	(ASTM D-5185-05 mod.)	8	8	5	6
Content in Ni (ppm)	(ASTM D-5185-05 mod.)	0	0	0	0
Content in Pb (ppm)	(ASTM D-5185-05 mod.)	1	1	3	4
Content in Si (ppm)	(ASTM D-5185-05 mod.)	0	6	4	4
Content in Sn (ppm)	(ASTM D-5185-05 mod.)	1	0	0	0
Content in V (ppm)	(ASTM D-5185-05 mod.)	0	0	0	0
Particles						
PQ index (Adim)	(PE-5024-Al)	3	89	8	12
Diagnosis						
Sample Diagnosis		Δ	X	Δ	√

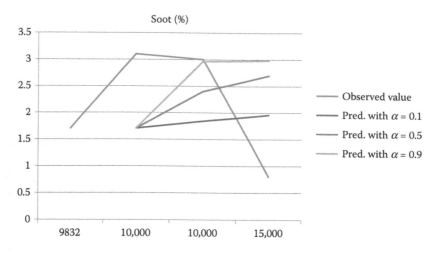

FIGURE 12.6
Soot analysis with forecasting.

TABLE 12.4

Soot Analysis with Forecasting

Soot (%)

Period	Observed Value	Pred. with $\alpha = 0.1$	Pred. with $\alpha = 0.5$	Pred. with $\alpha = 0.9$
9832	1.7			
10,000	3.1	1.70	1.70	1.70
10,000	3	1.84	2.40	2.96
15,000	0.8	1.96	2.70	2.98

Figure 12.6 shows an example of forecast, according to several values for smoothing parameter, for Soot variable, according to Table 12.4.

Table 12.5 shows an example of forecasting, according to several values for smoothing parameter, for Aluminum variable.

The control variable interval to take oil samples varies between 15,000 and 50,000 km.

The monitoring was done through periodic collection of oil samples from several selected vehicles. Data from older samples of the same vehicles were also used.

TABLE 12.5

Aluminum Analysis with Forecasting

Content in Al (ppm)

Period	Observed Value	Pred. with $\alpha = 0.1$	Pred. with $\alpha = 0.5$	Pred. with $\alpha = 0.9$
9832	2.00			
10,000	2.00	2.000	2.000	2.000
10,000	1.00	2.000	2.000	2.000
15,000	3.00	1.900	1.500	1.100

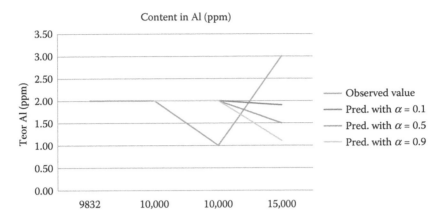

FIGURE 12.7
Aluminum analysis with forecasting.

The individual vehicle data that accompany each file of each oil sample are the following:

- Vehicle number
- Brand
- Model
- Type of vehicle
- Type of engine
- Kilometers traveled
- Date of the sample
- Sample submission date

Figure 12.7 shows an example of forecasting, according to several values for Aluminum variable.

Several variables were studied, although the project was focused only on the ones considered more important for the assessment of the oil degradation, namely the following:

- Soot (carbon matter)
- Viscosity
- Total base number (TBN)
- Wear and contamination metals
- Particles

Therefore, for the study of the variables used as reference, the limits suggested by the laboratory where they were processed were used. One of the variables considered most important and necessary to the study was soot or carboniferous matter (%).

A similar analysis is done for water content, nitration, oxidation, sulfidation, TBN, viscosity, chromium, copper, and silicon, among other contamination metals.

13

Dynamic Modeling

13.1 Background

A device can be modeled both by deterministic and stochastic models: a deterministic model is one that contains no random variable; that is, for a given set of known input data, there is a unique set of outputs. A stochastic model is one that is determined by a set of random variables indexed by parameters belonging to a given time interval; that is, if a variable corresponds to a real number that varies randomly, the stochastic model is a *temporal function* that varies randomly.

In terms of terology, it is of interest fundamentally to shape the behavior of assets, particularly in the areas of condition monitoring and fault diagnosis, with the stochastic models being the most adequate.

To support these maintenance areas, several mathematical tools from different knowledge areas can be used.

This chapter presents the general principles of operation of fault trees, Markov models and Petri networks, respectively, which are suitable for a large number of situations where it is necessary to model the operation of assets and, consequently, also analyze their faults.

13.2 Fault Trees

Fault tree analysis was developed by engineers at Bell Labs Telephone Company in the early 1960s.

Fault tree analysis belongs to the deductive methods used for the analysis of equipment and systems because they allow implementation of predictive analysis to identify faults.

The deductive approach begins with the definition of unwanted events, such as a failure or imagined or real accident in the case of an inquiry, allowing graphical organization, in a systematic way, of all known events that could contribute to or cause failures or undesired events. There is a large body of existing literature about this subject, for example, Olmos (1979) and Andrews and Dugan (1998).

13.2.1 Main Description of the Method

The purpose of fault tree analysis is the identification of combinations of occurrences in equipment or systems that could result in failure.

The main applications of this type of analysis are the following:

a. In the project phase—The fault tree is used to identify the modes of hidden faults that result from combinations of occurrences in equipment or systems, including human error in operation.

b. In the operation phase—The fault tree may include the features and operating procedures to study a device or system with the objective of identifying the combinations of potential events that may cause crashes.

The results of the analysis may be qualitative or quantitative. In the latter case, the failure probabilities of components have to be known, but fuzzy logic may also be used.

For the construction of a fault tree, the following are required (Figure 13.1):

a. Having full knowledge of the functioning and operation of equipment.

b. Having full knowledge of the failure modes of the components of the equipment and their effects on them. This information can be obtained through FMEA or FMECA analysis.

c. Sequencing the importance of failures can also be obtained through a GUT matrix, as will be seen later.

The result of the analysis through fault tree failure is a list of fault combinations in facilities and equipment, which is known as a minimum reduction set (MRS). Each set

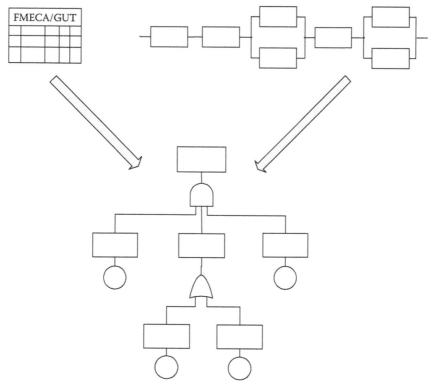

FIGURE 13.1
Construction of a fault tree.

corresponds to the lowest possible combination of events sufficient to cause a fault, since those occur simultaneously.

13.2.2 Logic Symbols Used in the Method

Analysis by fault trees is done using a graphical representation where interrelations are illustrated between the occurrences in equipment, including in its operation, that can result in failure. The symbols commonly used are illustrated below.

Logic gates:

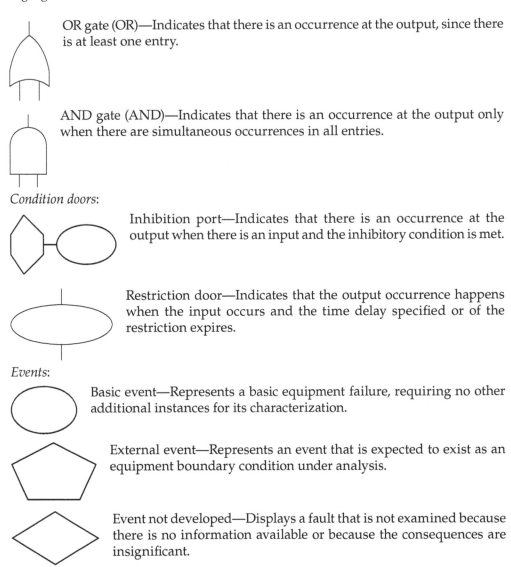

OR gate (OR)—Indicates that there is an occurrence at the output, since there is at least one entry.

AND gate (AND)—Indicates that there is an occurrence at the output only when there are simultaneous occurrences in all entries.

Condition doors:

Inhibition port—Indicates that there is an occurrence at the output when there is an input and the inhibitory condition is met.

Restriction door—Indicates that the output occurrence happens when the input occurs and the time delay specified or of the restriction expires.

Events:

Basic event—Represents a basic equipment failure, requiring no other additional instances for its characterization.

External event—Represents an event that is expected to exist as an equipment boundary condition under analysis.

Event not developed—Displays a fault that is not examined because there is no information available or because the consequences are insignificant.

State description:

Intermediate event and comments—Represents a failure as a result of the interaction of events that are developed through the logic inputs, as described above, or is used to make comments that are considered relevant.

Transfer symbols:

 Transfers—Indicates that the fault tree is developed on more than one page. The transfer symbols are identified by numbers or letters. The left symbol means an input transfer and right an output.

13.2.3 Types of Faults Analyzed

Equipment failures analyzed by fault trees can be grouped into three types:

1. Failures and primary defects
   - Correspond to equipment malfunctions that may occur in the environment and normal operating conditions for which they were designed. These failures are related to the intrinsic characteristics of the equipment and cannot be attributed to exogenous causes.
2. Faults and minor defects
   - Correspond to equipment malfunctions that may occur in environments and conditions for which they were not designed. These failures are attributed to exogenous causes.
3. Faults and defects of commands
   - Correspond to equipment failures that are related to the operation of control commands and operation.

13.2.4 Method Application

There are four steps to building a fault tree:

1. Problem definition
2. Construction of the fault tree
3. Fault tree analysis
4. Determination of the minimum reduction set

These steps are discussed in detail below.

1. *Problem definition*: The problem definition consists of the following steps:
   a. Set the main event
      - The main event is the most important aspect of the equipment definition. It corresponds to an undesirable fault that significantly affects the equipment's performance. Setting this occurrence ought to be done as accurately as possible and indicate what failed and when it happened.
   b. Set all related events
      - It is important to list all the related events with the main event considered in the analysis of the equipment, as well as the intermediate steps that relate

to other equipment. One way to define these events is through the analysis of their contribution to the development of the main event.

c. Set the equipment physical limits

   – The physical limits of equipment ought to be defined, which includes all events to be considered in the fault tree. One way to define the physical limits is by marking the process flowchart of which subequipment ought to be considered.

d. Set the resolution level of analysis

   – In addition to the physical limits of the equipment or system, the resolution level of analysis, which determines the level of detail in the equipment module disassembly for the analysis, must be set up. One factor to be considered in the resolution level of analysis to be used is the quantity of details available about the equipment failure. For this, first a FMECA analysis and/or GUT matrix analysis must be carried out.

e. Set other assumptions

   – Other assumptions must be established when necessary to describe the equipment in the most complete possible way, such as its mode of operation and its ability indexes C_{pk}, among others.

2. *Construction of a Fault Tree*: The construction of the fault tree starts with the main event and continues, level by level, until all events related to the main event have been developed to the most basic events.

   The analysis starts with the main event and, at the next level, it determines the immediate causes that give rise to the main event. Usually, these causes are not basic and intermediate, but causes that give rise to an additional development. If the root causes of the main event are immediately determined, it is because the problem is too easy to require the use of a fault tree analysis.

   The basic rules to be followed to build a fault tree are the following:

   a. Record the failure.

      – The event is described accurately within the relevant symbol. An additional report will be done indicating how, where, and when the fault happened. These reports must be the most complete possible and the analyst must resist to the temptation to shorten them.

   b. Assess the failure.

      – When the fault is assessed, one should ask if it happens for bad equipment in endogenous operation. If the answer is yes, the event is classified as a fault. If the answer is no, the event is classified as an exogenous fault.

   c. Establish the normal faults.

      – In normal operation, several failures usually happen that may be considered normal.

   d. Always complete all levels.

      – All the necessary entries for an event occurrence must be analyzed and recorded before moving to other event. The fault tree must be filled by levels, and it must be completed at each level before beginning the analysis of the next level.

    e. Connection of inputs for logic gates.

- The inputs must be defined as events that can cause failures and therefore must always be connected through a logic gate.

3. *Fault Tree Analysis*: A complete fault tree provides lots of useful information through the graphical and logical demonstration of the sequence of failures that may result in equipment failure. However, except for very simple fault trees, even a very experienced analyst cannot immediately identify all combinations that lead to equipment failure.

Fault trees can be solved by mathematical methods, such as Boolean algebra and matrix theory. Both methods give as a result the minimum reduction set that indicates the combinations of endogenous equipment failures that can result in major failures. MRS is useful to prioritize the ways in which the main failure may occur, allowing further quantification of their probability of occurrence.

The method to analyze the fault tree has four steps:

    a. To identify all inputs and basic events

    b. To simplify all entries to the basic events

    c. To remove duplicate events in the tree

    d. To suppress all series containing other series as subseries

The basic (or initial) event always corresponds to the first entry in the tree and ought to be clearly defined at the beginning of the analysis.

4. *Determination of Minimum Reduction Set*: The hierarchy of the MRS is the final step of the analytical procedures of the fault tree, and two factors must be considered:

    a. The first is of structural significance, being based on the number of basic events that are in each MRS. For example, an MRS with one event is more important than an MRS with two events; an MRS with two events is more important than one with three, and so on. This prioritization means that one event is more likely to occur than two, two are more likely than three, and so on.

    b. The second considers the hierarchy within each MRS series. The general rule governing this ranking is as follows:

       I. Human error

      II. Faults in active equipment

    III. Faults in passive equipment

Next, a simple example of application of fault trees is presented, based on the case of a failure in the main electricity supply, a public network, in a shopping center. There is an emergency generator driven by a diesel engine, which turns on when the outside power fails. The emergency generator feeds the emergency circuits, including emergency lighting and some circuits related to security.

The fault tree can be designed from the following characterization of the fault (Figure 13.2):

- Fault of electricity power supply to shopping center
  - Main event
    - Failure of main power source (public network)
  - Intermediate events
    - Emergency generator failure

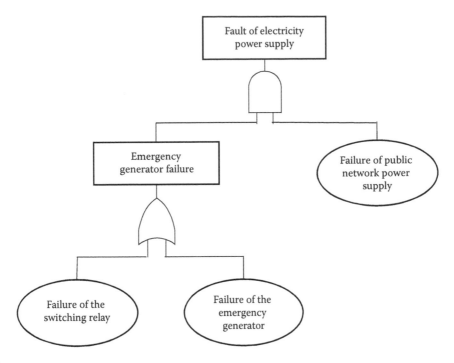

FIGURE 13.2
Fault tree related to power failure in a shopping center.

- Basic events
  - Failure of main power source (public network)
  - Failure of relay that switches between the public network and the emergency generator
  - Failure of the emergency generator to power on

13.3 Markov Chains

A Markov chain is a special case of a stochastic process with discrete states (the parameter, which in general is the time, may be discrete or continuous—although the most usual approach is the discrete state) having the Markovian property, so called in honor of the mathematician Andrei Andreyevich Markov. There is extensive existing literature on this subject (Norris, 1998; Basharin et al., 2004; Wai-Ki Ching, 2009).

13.3.1 Main Description of the Method

A Markov chain is a particular case of a stochastic process with discrete states, having the Markovian property, also called Markov memory, which means that the previous states are irrelevant for predicting the following states, since the present state is known.

In the case of discrete Markov chains, the time instants in which the transition from one state to another occurs can only have integer values: 0, 1, 2..., n.

The processes must remain in a given state for a given time, which ought to be geometrically distributed.

A Markov chain is a sequence $X_1, X_2..., X_i$ of random variables. The set of values that these variables can assume, s_i, is called the *state space*, where X_{t+1} designates the state of the process at time $t + 1$. If the conditional probability distribution of X_{t+1} in past states is a function only of X_t, then:

$$P(X_{t+1} = s_{t+1} | X_0 = s_0, X_1 = s_1, ..., X_{t-1} = s_{t-1}, ..., X_t = s_1)$$
$$= P(X_{t+1} = s_{t+1} | X_t = s_t) = P_{ij} \geq 0 \tag{13.1}$$

where s_{t+1} is a state of the process. The preceding equation defines the *Markov property* for $t = 0, 1,...$ for any sequence of s_i.

This property, as defined above, is equivalent to stating that the conditional probability of any future event, given any past event and the present state $X_{t=i}$, is independent of the past event and depends only on the present state of the process.

The conditional probabilities $P(X_{t+1} = s_{t+1} | X_t = s_t)$ are called transition probabilities.

If, for each s_t and s_{t+1}, $P(X_{t+1} = s_{t+1} | X_t = s_t) = P(X_1 = s_{t+1} | X_0 = s_t)$, then the transition probabilities (of one step) are called homogeneous, or *stationary*, and are typically denoted by P_{ij} which means that the stationary transition probabilities do not change with time.

The existence of stationary transition probabilities (of one step) also implies that, for each s_t, s_{t+1}, and n ($n = 1, 2,...$):

$$P(X_{i+n} = s_{t+1} | X_t = s_t) = P(X_n = s_{t+1} | X_0 = s_t) \tag{13.2}$$

Theorem: The transition probabilities in n steps of a Markov chain satisfy

$$P_{ij}^{(n)} = \sum_{k=0}^{\infty} P_{ik} P_{kj}^{n-1} \tag{13.3}$$

where

$$P_{ij}^{(0)} = \begin{cases} 1, & \text{if } i=j \\ 0, & \text{if } i \neq j \end{cases} \tag{13.4}$$

From the iteration of Equation (13.3), one obtains:

$$P^{(n)} = P * P * ... * P = P^n \tag{13.5}$$

...n times....

These conditional probabilities are denoted by $p^{(n)}$ and are called transition probabilities of the n step. Thus, $p^{(n)}$ is the conditional probability for the random variable X, starting at state s_t. It will be in state s_{t+1} after n steps (time units).

The conditional probabilities $P_{ij}^{(n)}$ must satisfy the following properties:

$$P_{ij}^{(n)} \geq 0, \text{ for all } i \text{ and } j, \text{ and } n = 0, 1, 2,...$$

$$\sum_{j=0}^{M} P_{ij}^{(n)} = 1, \text{ for all } i \text{ and } j, \text{ and } n = 0, 1, 2,... \tag{13.6}$$

A good approach to representing these probabilities is the following matrix:

$$P^{(n)} = \begin{array}{c|ccc} State & 0 & 1 & M \\ \hline 0 & p_{00}^{(n)} & \cdots & p_{0M}^{(n)} \\ \vdots & \vdots & \vdots & \vdots \\ M & p_{M0}^{(n)} & \cdots & p_{MM}^{(n)} \end{array} \text{, for } n = 0,1,2,\ldots \tag{13.7}$$

Given the above, it can be defined synthetically that a stochastic process $\{X_t\}$ ($t = 0, 1,\ldots$) is a *finite-state Markov chain* if the following conditions exist (Figure 13.3):

- It has a finite number of states.
- It has a Markov probability.
- It has stationary transition probabilities.
- It has a set of initial probabilities $P\{X_0 = i\}$ for all i.

To show the application of the method, the following example is presented:

- Assume that the waste gases of a given heat engine may be in one of the two following states:
  - The emissions are below the environmentally acceptable limits.
  - The emissions are above the environmentally acceptable limits.

According to readings made during maintenance interventions in the exhaust system, the following can be verified:

- If the technician measures emissions below the environmentally acceptable limits, there is a 90% chance this technician will again measure these emissions below environmentally acceptable limits.
- If the technician measures emissions above the environmentally acceptable limits, there is an 80% chance this technician will again measure these emissions above the environmentally acceptable limits.

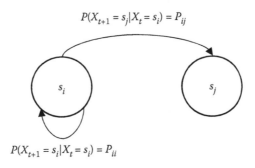

FIGURE 13.3
Stationary transition probability.

According to the above, the following questions are posed:

1. If the technician currently measures emissions above the environmentally acceptable limit, what is the probability that he or she will measure emissions below the environmentally acceptable limit in the second measurement?

2. If a technician currently measures emissions below the environmentally acceptable limit, what is the probability that he or she will measure emissions below the environmentally acceptable limits in a third measurement?

The analysis of the evolution of states in the perspective under analysis can be interpreted through a Markov chain of two states, described as:

- State 1—The technician measured emissions below the environmentally acceptable limits the last time (S_1).

- State 2—The technician measured emissions above the environmentally acceptable limits the last time (S_2).

Setting the values of the states as:

X_0—Measured value of emissions in the present state

X_n—Emission value at the nth future state

The sequence $X_0, X_1, ...$ can be described by a Markov chain described by the following matrix of states, as is shown in Figure 13.4:

$$P^{(n)} = \begin{array}{c} \\ 1 \\ 2 \end{array} \begin{array}{cc} State\ 1 & 2 \\ \hline 0.9 & 0.1 \\ 0.2 & 0.8 \end{array}$$

Let's answer the first question:

1. If the technician currently measures emissions above the environmentally acceptable limits, what is the probability that he or she will measure emissions below the environmentally acceptable limits in the second measurement?

The goal is to know the following:

$$P(X_2 = S_1 | X_0 = S_2) = p_{21}^{(2)}$$

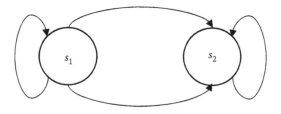

FIGURE 13.4
State diagram of effluents.

And applying Equation (13.3) gives

$$P_{21}^2 = \begin{bmatrix} 0,9 & 0.1 \\ 0.2 & 0.8 \end{bmatrix}\begin{bmatrix} 0,9 & 0.1 \\ 0.2 & 0.8 \end{bmatrix} = \begin{bmatrix} 0,83 & 0.17 \\ 0.34 & 0.66 \end{bmatrix}$$

The resulting final probability of the previous matrix and diagram can be described as follows (Figure 13.5):

- P_{21}^2 = (probability that the next state is above the environmentally acceptable limits and the next state is above the environmentally acceptable limits) + (probability that the next state is below the environmentally acceptable limits and the next state is above the environmentally acceptable limits) =

$$P_{21}^2 = p_{21}p_{11} + p_{22}p_{21} = (0.2)(0.9) + (0.8)(0.2) = \mathbf{0.34}$$

Let's answer the second question:

2. If a technician currently measures emissions below the environmentally acceptable limits, what is the probability that he or she will measure emissions below the environmentally acceptable limits in a third measurement?

This question aims to find the following:

$$P(X_3 = S_1 | X_0 = S_1) = p_{11}^{(3)}$$

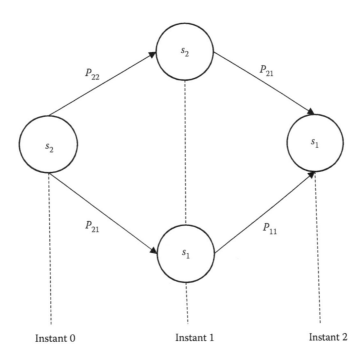

FIGURE 13.5
Transition diagram between states—first question.

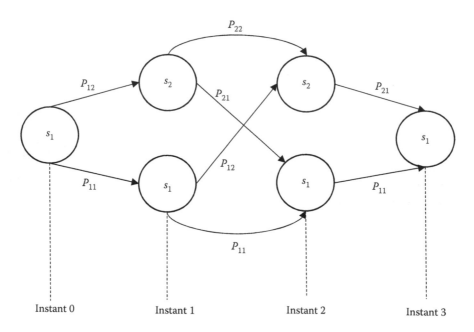

FIGURE 13.6
Transition diagram between states—second question.

$$P_{11}^3 = P_{12}P_{21}^2 = \begin{bmatrix} 0,9 & 0.1 \\ 0.2 & 0.8 \end{bmatrix} \begin{bmatrix} 0,83 & 0.17 \\ 0.34 & 0.66 \end{bmatrix} = \begin{bmatrix} 0,781 & 0.219 \\ 0.438 & 0.562 \end{bmatrix}$$

The resulting final probability of the previous matrix and diagram is the following (Figure 13.6):

$$P_{11}^3 = p_{12}p_{22}p_{21} + p_{12}p_{21}p_{11} + p_{11}p_{12}p_{21} + p_{11}p_{11}p_{11}$$
$$= (0.1)(0.8)(0.2) + (0.1)(0.2)(0.9) + (0.9)(0.1)(0.2) + (0.9)(0.9)(0.9) = 0.781$$

13.4 Hidden Markov Models

Another aspect of Markov chains particularly relevant for equipment condition monitoring can be managed by hidden Markov models (HMMs), which are summarized in Figure 13.7, where x_i corresponds to the variable States, y_i, to Possible Observations, a_{ij} to Probabilities of State Transitions, and *bij* to Exit Probabilities.

An HMM is a statistical model in which it is assumed that a system is modeled through unknown parameters. The challenge is to determine the hidden parameters from observed data.

In an HMM, the state is not directly visible; only the variables influenced by the state are visible, and for each of these, there is a probability distribution among the possible outputs. The succession of outputs generated by an HMM gives information about the succession of states.

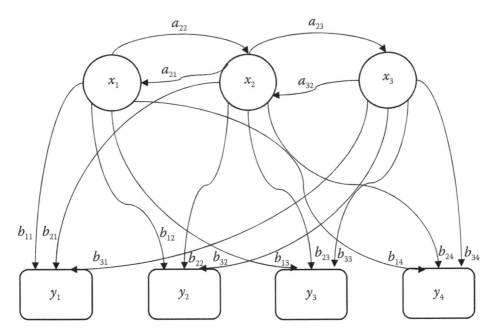

FIGURE 13.7
Example of probabilistic parameters of a hidden Markov model.

This subject will not be developed in this book, so the reader is referred to the specialized references (Rabiner, 1989; Cappé et al., 2010).

13.5 Petri Networks

Petri networks (PNs) come from the doctoral thesis of Carl Adam Petri, presented at the University of Bonn in 1962. Its original purpose was to model systems with concurrent components.

According to Heuser (1991), the first Petri network applications were in 1968, in the United States, by the project Information System Theory of Applied Data Research, Inc. (ADR). ADR was a large-scale software sales company from the 1960s to the mid-1980s. An important part of the initial theory, notation, and representation of Petri networks was developed in this project and published in its final report. This work showed the potential application of these networks in analysis and systems modeling with competing components.

In the 1970s, development of the Petri network theory and the expansion of its scope were made relevant. Earlier in that decade, the Petri work caught the attention of members of Project MAC (Multiple Access Computer and Machine-Aided Cognition) at the Massachusetts Institute of Technology (MIT). The group of computer structures in this project, under the direction of Prof. Jack B. Dennis, made an important contribution to the development of Petri nets through their research and publications, including reports and PhD theses.

About this subject, see, for example, Wang (1998), Girault and Valk (2002), Diaz (2009), Popova-Zeugmann (2013), and Reisig (2013).

13.5.1 Main Description of the Method

According to Marranghello (2005), there are three basic elements that constitute the topological structure of Petri nets:

1. States (E)
   a. They are used to model the passive components of the system; that is, they correspond to their state variables to form a set:

   $$E = \{e_1, e_2, \ldots, e_n\}.$$

2. Actions (A)
   b. They are used to model the active components of the system, that is, the events that lead the system from one state to another, forming a set:

   $$A = \{a_1, a_2, \ldots, a_m\}.$$

3. Flow ratio (F)
   c. It is used to specify how a transformation happens from one state to another by the occurrence of actions in the system. This relationship is represented by a set:

   $$F = \{(x, y) \in E \times A \cup A \times E\}.$$

At this point, it is important to define the characteristics of Petri nets that use graph theory, which are highlighted in the following three characteristics:

1. Two-way graph—The graph whose vertices can be divided into two sets in which there are no edges between vertices of the same set—for a graph to be two way, it cannot contain odd-length loops.
2. Directed graph or digraph—A graph in which all edges are directed.
3. Connected graph—A graph in which it is possible to establish a path from any edge to any other edge of the graph.

On graph theory, many references may be consulted (Trudeau, 1994; Bondy and Murty, 2010; Deo, 2016).

The topological structure of a Petri net is given by the triple $R = (E, A, F)$, which defines a two-way, directed, connected graph, with the following characteristics:

a. $E \cup A \neq \varnothing$—The graph is not empty and has no isolated elements.
b. $E \cap A \neq \varnothing$—The sets of states and actions are disjoint.
c. $F \subseteq (E \times A) \cup (A \times E)$—The flow ratio is a set in the universe of states and actions, identifying the neighborhood relationship between these entities.
d. The domain of the flow ratio is given by

$$D(F) = \{x \in E \cup A \mid \forall\, (x, y) \in F \rightarrow \exists\, y \in E \cup A\};$$

e. The codomain of the flow ratio is given by

$$CD(F) = \{y \in E \cup A \mid \forall\,(x, y) \in F \rightarrow \exists\, x \in E \cup A\};$$

f. $D(F) \cup CD(F) = E \cup A$—The union of the domain and codomain of the flux relation corresponds to the universe of states and actions of the net.

Graphical notations commonly used for the representation of the elements of Petri nets are as follows:

- States (E) are represented by ellipses or circles.

- Actions (A) are represented by rectangles or squares.

- The flow ratio (F) is represented by arrows.

For example, a Petri net whose algebraic representation is given by:

$R = (E, A, F)$
$E = \{e_1, e_2, e_3, e_4, e_5\}$
$A = \{a_1, a_2, a_3, a_4\}$
$F = \{(e_1,a_2),\,(e_2,a_2),\,(e_3,a_1),\,(e_5,a_4),\,(e_4,a_3),\,(a_2,e_3),\,(a_3,e_1),\,(a_1,e_2),\,(a_4,e_4),\,(a_1,e_5)\}$ is represented graphically as illustrated in Figure 13.8.

Petri nets are supported by the following principles:

1. *General concepts*
   a. States and actions are concomitantly interdependent and distinct notions.
   b. States and actions are distributed entities.
   c. A *case* (*C*) is a subset of states distributed throughout the network and satisfied simultaneously.

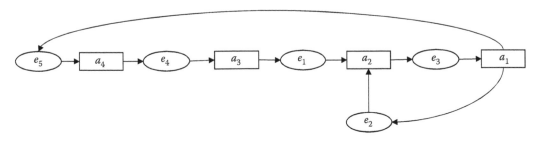

FIGURE 13.8
Graphical representation of a Petri net.

d. A step (P) is a subset of actions distributed across the network that may occur simultaneously.

e. The changes caused by an action are fixed and independent of the case in which they occur.

f. The occurrence of a state is graphically represented by a mark (•) placed within the corresponding ellipse.

g. Given $\delta \in \Delta = E \cup A$ (δ is a state or action of the network) and a network element $R = (E, A, F)$, its input elements are given by its preset, represented by $^\bullet\delta = \{x \in \Delta | (x, \delta) \in F\}$, and its output elements are given by its postset, represented by $\delta^\bullet = \{y \in \Delta | (\delta, y) \in F\}$.

h. An action can occur at a particular time if and only if the changes caused by the action were possible in the case considered.

i. A net is said to be pure if and only if $\forall\ \delta \in \Delta,\ ^\bullet\delta \cap \delta^\bullet = \varnothing$; that is, if it does not contain recursions, as shown in Figure 13.9.

2. *Constraints*

a. It can be said that an action a may occur in a case C if and only if all preconditions of a were contained in C and no one postcondition of a was satisfied in C.

$$C\,[a > \ \leftrightarrow\ \exists^\bullet a \subseteq C \wedge a^\bullet \cap C = \varnothing$$

b. By extension, a step P may occur in a case C if and only if all the elementary actions of P can occur individually in C without causing interference with each other.

$$P = \{a_i | a_i \in A \wedge P \subseteq A\} \rightarrow C\,[P > \ \leftrightarrow\ \forall a_i \in P, C\,[a>$$

3. *Effects of occurrences*

a. When an action occurs in C, its preconditions are no longer met, and their postconditions happen, remaining unchanged in the rest of the case.

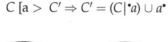

$$C\,[a > \ C' \Rightarrow C' = (C | ^\bullet a) \cup a^\bullet$$

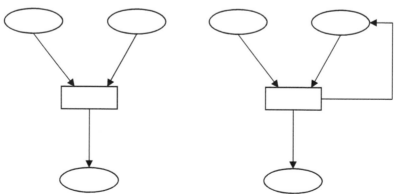

FIGURE 13.9
Example of a pure Petri net (left) and impure (right).

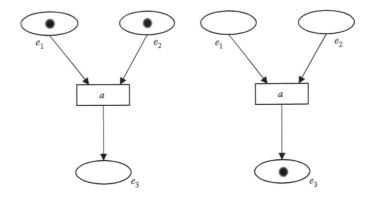

FIGURE 13.10
Effect of occurrence of an action [(a) before; (b) after].

b. By extension, the result of the occurrence of a step P in case C is the sum of the individual results of occurrences of the elementary actions of the step considered in the case.

$$P = \{a_i | a_i \in A \land P \subseteq A\} \rightarrow C \, [P > \, C' \Rightarrow C' = (C|\!^\bullet P) \cup P^\bullet$$

c. Assume an action a with preconditions e_1 and e_2 and postcondition e_3. Considering the enabled action in this case, the constraint of action a is represented by the marks in e_1 and e_2 (left side [a] of Figure 13.10) and the effect of their occurrence by the removal of the marks in e_1 and e_2 and the addition of a mark in e_3 (right side [b] of Figure 13.10).

$$^\bullet a = C = \{e_1, e_2\} \land a^\bullet = \{e_3\} \rightarrow C' = C/\{e_1, e_2\} \cup \{e_3\} = \{e_3\}$$

13.5.2 Elementary Petri Networks

Elementary Petri nets synthesize several evolutions of the original model proposed by Petri in 1962, known in the literature as *classical* or *condition/event* networks, preserving its original features.

An elementary Petri net is a quadruple $RE = (E, A, F, C_{in})$, where:

1. $R = (E, A, F)$ is the home network, that is, representing the static topological structure of the modeled system, keeping the rating, terminology, and initial concepts of networks.

2. C_{in} is an initial event, which represents the initial dynamic state of the system immediately before its start.

The graphical representation of an elementary network consists of the initial graphic notation of the network plus the markings. This association of a network topological structure with a set of marks is called a scheduled network.

The actions of an elementary network have several ways to relate among themselves in a case C, namely the following ways:

1. Sequence
2. Conflict

3. Competition

4. Confusion

5. State space

1. Sequence
   - It is said that a_1 and a_2 form a sequence (S) in a case C_1 if and only if a_1 may occur in C_1 and a_2 cannot. However, after the occurrence of a_1, a_2 is able to happen.

2. Conflict
   - It is said that a_1 and a_2 are in conflict in a case C if and only if a_1 and a_2 may occur individually in case C, but cannot occur simultaneously. Therefore, $\{a_1, a_2\}$ does not correspond to a step in case C. Because it is not possible to say which of the actions occur first, before the system is running, it can be said that the basic network is not deterministic.

3. Competition
   - It is said that a_1 and a_2 can happen in competition in a case C if and only if they do not suffer mutual interference. There is no specified order for the occurrence of actions making up the step enabled in C. Thus, the occurrence of actions and the states arising therefrom will be partially ordered, where elementary networks can display nonsequential behavior.

4. Confusion
   - This is a situation resulting from the mix of competition and conflict. Given a, an action of action space C, the set of conflicts in a of C, named $cfl(a,C)$, is the set of actions a' with occurrences in C that do not form a step with a; that is, $cfl(a,C) = \{a' \in A/C\ [a' > \land (C\ [\{a, a'\}>)]\}$. Therefore, for any two actions a_1 and a_2 with occurrences in C_1, the triple (C_1, a_1, a_2) is a confusion in C_1 if and only if the sets of conflict of a_1 in C_1 and in C_2 were different and the execution of a_2 in C_1 will result in C_2.

5. State space
   - The state space of an elementary network (C_{RE}) is the set of all cases of the network that occur during system execution ($R_{CE} = [C_{in}>)$. This space is used in the analysis of the network properties.

Elementary networks primarily allow the study of competition and the theoretical development of Petri nets. However, this approach is not easy for practical use because the models developed with these networks increase rapidly to a very large number of elements, even in a simple modeling system. An attempt to overcome this difficulty, since the 1970s, has been by extensions of classical networks, namely the following:

- Networks of place/transition
- High-level networks
- Other extensions of Petri nets:
  - Hierarchical networks
  - Timed networks:
    - Deterministic timed networks
    - Stochastic timed networks

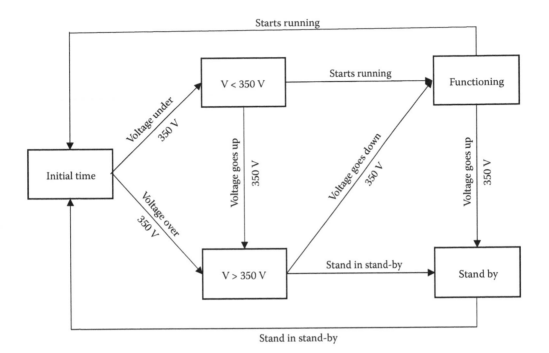

FIGURE 13.11
State diagram.

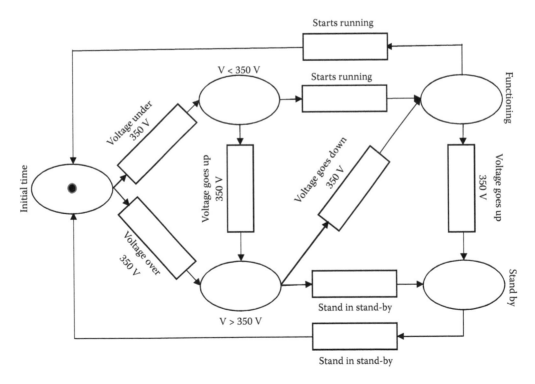

FIGURE 13.12
Petri network.

EXAMPLE

Here is presented an example that corresponds to an emergency generator that starts functioning when the external power supply voltage falls below a certain value of nominal voltage. In the example, the default value for the emergency generator to turn on is 350 V; that is, when the value of the external power supply voltage falls below this value, the generator starts working, turning off when the value of external power supply is above 350 V.

To solve the problem, the following situations are assumed for the emergency generator:

- The generator can be in two possible operating states: in standby and running (generating electricity).
- There are two situations that give rise to those states: voltage above 350 V (>350 V) and below this value (<350 V).
- Other possible states, such as fault, are not considered.

Figures 13.11 and 13.12 illustrate the state diagram and Petri net for the previous situations, respectively.

13.6 A Case Study

The example presented here refers to an electrical-source energy power system in a hospital in Europe and is also used in Chapter 15. The system has several levels of redundancy and is controlled by an intelligent load switch, as is shown in Figure 13.13.

Figure 13.14 shows the Petri network that models the electrical power system.

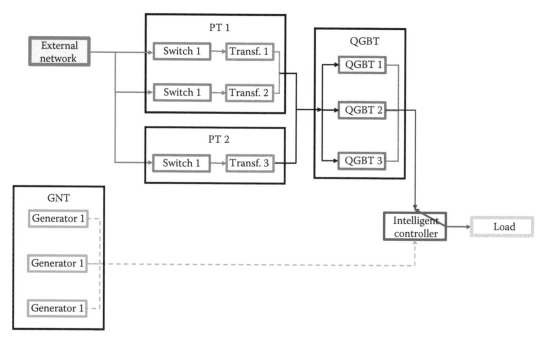

FIGURE 13.13
Electrical power system in an hospital.

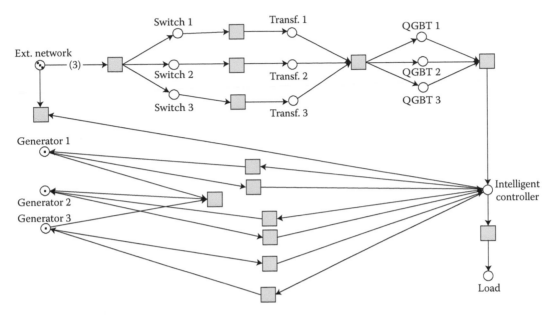

FIGURE 13.14
Petri network that models the electrical power system.

14

Three-Dimensional Systems

14.1 Background

Today, 3D modeling is a usual technique for equipment design, redesign, renewal, and maintenance. In fact, equipment can be fully designed through 3D software tools with high precision, including simulating material behavior. Three-dimensional modeling can be static or dynamic, with the latter the most interesting one because the user and/or technician can easily understand the equipment's functioning and make its maintenance easier.

If the equipment has a 3D model when it is purchased, this information must be introduced to the CMMS and be accessible to the technicians when they do maintenance interventions and or other types of activities like the above mentioned.

However, until now, it has not been usual for the manufacturer to supply a 3D model of the equipment. But, if necessary, in order to facilitate maintenance interventions and similar activities, it is possible to make a 3D model of equipment at any time. This approach may be relevant in the following situations:

- Renewal
- Maintenance interventions with high levels of difficulty
- Markerless augmented reality

There are a lot of books and papers about this subject (May and Christensen, 2015; Bethune, 2016; Giesecke et al., 2016; Shih, 2016; Thilakanathan, 2016).

14.2 Three-Dimensional Models for Maintenance

Maintenance interventions are usually classified into five levels, from one to five, with the first the easiest and the fifth the most complex. However, for the same maintenance level, the difficulty of maintenance intervention varies from equipment to equipment. Obviously, for the same level, a very complex piece of equipment implies much more difficulty than a simple one. This implies that it may not be necessary to always use the same resources for all equipment, even at the same maintenance level.

When there aren't 3D models from the manufacturer, it is necessary to implement them through a specialist, which is very expensive. Then, it is necessary to evaluate profits versus expenses. But, when implemented, 3D models are extremely important to aid maintenance interventions, which are maximized if the user can manipulate them dynamically.

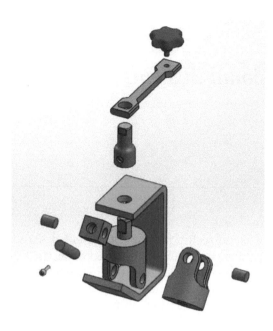

FIGURE 14.1
3D model.

A working order for each specific intervention includes procedures, materials, tools, and, obviously, human resources.

When technicians perform interventions, they use their expertise to conduct equipment maintenance. However, even following the procedures and having the correct spare parts to use, a lot of doubts may appear during the intervention.

If a technician has a 3D model and the WO makes a correct correspondence between procedures, modules, and components of the 3D model, the intervention becomes easier and quicker and, finally, has higher quality. A 3D model corresponds to a picture such as that illustrated in Figure 14.1.

If the models are animated, the technicians can access each component in more detail and evaluate the interconnections, movements, and other details that can help in the intervention. This also helps to identify the sequence of operations and the details of each one. The equipment assembly and disassembly operations become much easier to do and errors are extremely diminished.

Finally, if these operations are accompanied by an augmented reality tool, the maintenance activity reaches an historical level that results, in the last instance, in increased asset reliability and availability.

14.3 Three-Dimensional Models and Maintenance Planning

Maintenance in general and maintenance planning in particular can use all the advantages of time and quality in maintenance interventions, as mentioned in the preceding section.

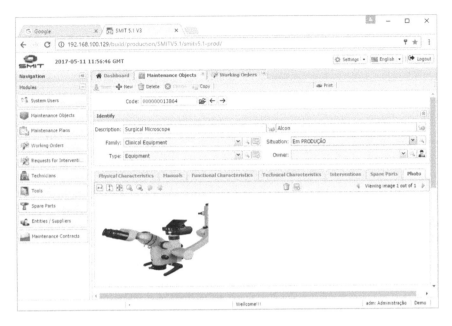

FIGURE 14.2
3D model embedded in the form of an asset.

Planned interventions may take great advantage of 3D modeling for many reasons, namely the following: the selection of the 3D model necessary for each intervention can be made with or without animation, and then it will always be used in similar interventions, because the planning intervention content repeats forever; the connections between each spare part code and the 3D models drastically simplify the operations and diminish errors, and the sequence of operations becomes unmistakable and, as a consequence, the intervention time minimal.

Figures 14.2 and 14.3 show two forms: the first with characterization of the physical asset with an example of a 3D model and the second with a working order for that asset with the same 3D model, showing a possible interaction between these two modules.

The use of this approach has many additional advantages, namely the following:

- Even if the technician who usually does the intervention and is more experienced in the equipment is absent, it is easier for any other technician of the same specialty to replace the first and perform the intervention with the same quality;

- The 3D models are repeated, intervention after intervention, for the same maintenance plan, which permits much more value for the money.

14.4 Three-Dimensional Models and Fault Diagnosis

In the preceding section, it was emphasized that, for maintenance planning, 3D modeling has great potential. This section takes a similar approach, but for fault diagnosis, in which situation the quality increase and the decrease in time spent in maintenance

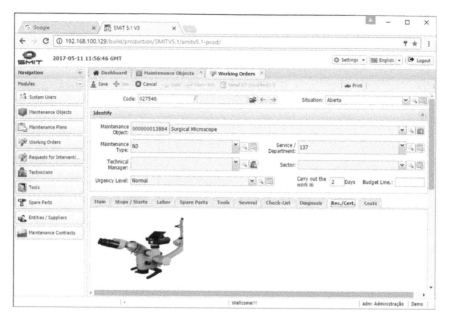

FIGURE 14.3
3D model embedded in a WO of an asset.

intervention may represent increasing profits even when compared to planned interventions using 3D models.

Traditional systems are based on a hierarchical structure, like the one inserted in SMIT, referred to in Chapter 7, or the approaches described in Chapter 8.

However, these systems, friendly as they are, have always some complexity. Three-dimensional modeling may represent the difference between an interface of a unfriendly fault diagnosis system based on artificial intelligence and an interactive system based on AI and a 3D system, where the proposed solution for each fault is followed by a graphical 3D explanation about what to do.

In fact, each fault implies a specific diagnosis that ought to be aided by an AI system. These tools are both based on historical data, if it exists and is part of the technician's knowledge through, for example, fuzzy tools.

This is an emergent research area, where there is some research done, but a lot still to do. Some of many research works about this subject are referenced next.

Houten and Kimura (2000) present a work about digital (geometric) product models that can be used for maintainability analysis and maintenance planning. Some examples are product life cycle simulation, deterioration analysis, FMEA analysis, product model-based monitoring, failure diagnosis, disassembly analysis for repair and replacement, and maintenance ergonomic analysis.

Ciang et al. (2008) present a review of damage detection methods for structural health monitoring of a wind turbine system, which is another strategic area where 3D modeling and fault diagnosis ought to work together.

Saeed et al. (2013) discuss condition monitoring and fault diagnosis in a Francis turbine based on integration of numerical modeling with several different artificial intelligence techniques.

14.5 Three-Dimensional Models and Robots

Nowadays, it is possible to conjugate 3D models, robots, and AI in a synergetic way, such as for critical facilities like nuclear power plants, oil platforms, and so on. There are many applications that use these concepts, and, even though there are many weak points in this knowledge area to fix, in fact, these new tools represent a considerable advance in the maintenance area.

Pouliot and Montambault (2008) present a design of a teleoperated robot called LineScout that has the capacity to cross obstacles found on a line. Many 3D models aid the design and implementation of the robot prototype, like the following: LineScout gripper and roller, LineScout flipping arm, and LineScout Center Frame.

Aksel et al. (2010) present a system solution, including components such as a 3D robot vision system, a robot tool, and a control architecture for remote inspection and maintenance interventions on processes similar to those on topside oil platforms. To implement the solution, the researchers use 3D modeling in many parts of the project. For example, they simulate, through 3D models, parts of a production process on a real oil platform, and two robot manipulators are used for inspection and maintenance tasks on the process structure using some tools and sensors. They also use 3D models for many others parts of the project, like detecting positions for sensors, among others.

Many other research projects on the use of robots for fault diagnosis and maintenance operations have been presented, including some using virtual reality techniques. One of those projects includes a system for collecting vibration data with the use of a remote robot controlled by a virtually designed model. With this model, the researchers manipulated both the equipment to be checked and the robot. They used 3D models to identify the data collection points in order to ensure good exactitude in the data collection.

Wang et al. (2014) developed a 3D model-driven remote robotic assembly system. They constructed 3D models at runtime that can represent unknown geometries on the robot side. Guided by the 3D models, a remote operator can manipulate a real robot instantly through the Internet for remote assembly operations.

The world of 3D modeling and robots is innovating year after year, day after day. These are knowledge areas that are becoming more and more a part of physical asset maintenance structural tools (Bellamine et al., 2002).

14.6 Software Tools

There are several software tools to develop 3D models. Some of the most used are the following:

- AutoCAD (http://www.autodesk.com/products/autocad/overview)
  - AutoCAD is a 2D and 3D computer-aided drafting software application used to assist in the preparation of blueprints and other engineering plans.
- Inventor (http://www.autodesk.com/products/inventor/overview)
  - Inventor permits the design of 3D objects, including freeform, direct, and parametric modeling options, design automation tools, and advanced simulation and visualization tools.

- Fusion 360 (http://www.autodesk.com/products/fusion-360/overview)
  - Some of features of Fusion 360 are the following:
    - Permits parametric modeling
    - Permits integration of CAM 360
    - Introduction of Publisher 360
    - Introduction of 2D drawings
    - Expanded collaboration and data sharing
    - Expanded APIs for developing custom macros and functions
    - Expanded access to 3D printing
- Revit (http://www.autodesk.com/products/revit-family/overview)
  - Revit allows users to design a building and structure and its components in 3D, annotate the model with 2D drafting elements, and access building information from the building model's database.
- Blender (https://www.blender.org/)
  - Blender is a free and open-source 3D creation suite, supporting the entirety of the 3D pipeline, that is, modeling, rigging, animation, simulation, rendering, compositing, and motion tracking, even video editing and game creation.

There are many other software tools for 3D modeling, with different features among them. The user may choose the best solution for his or her specific problem. However, this option must be carefully studied before the final decision because of the high investment in training and because this is usually a long-term option.

14.7 A Case Study

Because this knowledge area has had enormous developments and will continue at the same pace in the next years, the example presented here is based on a paper by Moczulski et al. (2013). Because of that, only a synthesis will be done of the interesting application called interactive education of engineers in the field of fault diagnosis and fault-tolerant control.

The goal is to create a 3D virtual reality model of the second stage of a water filtration system, together with its simulation model, to be used as a learning tool for stimulating education in the domain of advanced control theory in a university course. Some applications of the tool for fault diagnosis are given, and this approach is discussed.

One example given is a project related to a fossil-fuel power station. The application consists of three logical parts:

1. The first is a virtual walkthrough of a power station, allowing observation of the stages of electric power generation. The objective is that the user understand the principle of operation of the power station.
2. The second presents selected issues of machines and their maintenance, particularly shaft alignment of the turbo generator, rotor balancing, and others.

3. The third corresponds to the diagnostics of machines and industrial processes. Here, the user can inspect the turbine, blades, bondage, and wheels through endoscopic technology.

This paper adds a case study for the subject under discussion: analyzing the fault detection and isolation for the water filtration system and evaluating fault-tolerant control of the plant.

The fault models are the usual ones used for diagnosis and are based on the simulations made in 3D models.

15

Reliability

15.1 Background

This chapter discusses reliability and related concepts, such as quality, maintainability, and dependability, as well as the main statistical distributions applicable to the evaluation of reliability parameters.

Reliability is a complex discipline and requires, in an exhaustive treatment, a specific monograph. However, due to its importance to physical asset maintenance, this chapter addresses some relevant aspects of this knowledge area, including the synthesis of the most common statistical distributions used for reliability analysis.

15.2 Reliability Concept

Reliability is the probability that a device or component of a system will operate within the defined quality parameters for a given period of time under pre-established operating conditions. The ending of its operation under these conditions is called failure.

The concept of reliability, according to EN 13306, is defined as the "ability of a good to fulfill a function required under certain conditions, during a given time interval." This standard also presents a footnote about this concept, which says that "the term" reliability "is also used as a measure of reliability performance and may still be defined as a probability."

Intrinsically linked to the preceding is the maintainability concept, which, according to EN 13306, is defined as the "ability of an item under given conditions of use, to be retained in, or restored to, a state in which it can perform a required function, when maintenance is performed under given conditions and using stated procedures and resources."

It is from this perspective that it is imperative to consider the maintainability concept from the decision time to making the investment in physical assets, specifically taking into consideration the following criteria:

1. About the design
   a. Standardization of facilities, equipment, and their components
   b. Modularity
   c. Indicators of wear, failure, and reference limits
   d. Counters of use
   e. Known technology

     f. Ease and time of diagnosis and failure resolution

     g. Ease and time of disassembly and reassembly

     h. Reset regulations

2. About user management

     a. Standardization of facilities and equipment

     b. Commissioning of facilities and equipment

     c. Ability to perform and manage maintenance

3. About documentation

     a. Quality of the technical documentation

     b. Standards for setup, maintenance, and use

4. About after-sales services from the supplier

     a. Evolution of facilities and equipment models over time

     b. Effectiveness and seriousness of after-sales service

     c. Acquisition of spare parts

     d. Guarantee of continuity of the supplies

Another important concept related to reliability is dependability, which, according to EN 13306, is defined as the "collective item used to describe the availability and its influencing factors: reliability, maintainability and maintenance supportability." The same norm adds an additional note: "Dependability is used only for general descriptions in non-quantitative terms."

The main attributes of dependability are the following:

- Reliability—Measure of the correct continuity of service (dependability in relation to the continuity of service)
- Availability—Measurement of the correct service delivery in relation to changes between correct and incorrect service (dependability regarding readiness to use)
- Recovery—Measurement of service delivery, correct or incorrect, after a noncatastrophic failure (dependability in relation to nonoccurrence of catastrophic failures)
- Security—Dependability regarding the prevention of unauthorized access and or use of information
- Robustness—The degree of confidence about proper functioning of a system or a component in the presence of invalid inputs or stress conditions of the surrounding environment
- Sustainability—The ability to maintain an industrial process using only the indispensable resources (materials and energy)

The foregoing concepts are fundamental at any point in the life cycle of a physical asset. However, it is essential to have them present from the time of its acquisition in order to evaluate the asset's availability, the necessary resources to maintain it, and the respective costs and, consequently, to support the decision about its most appropriate maintenance policy.

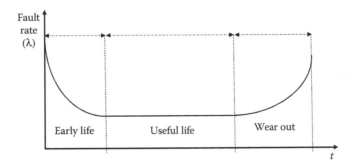

FIGURE 15.1
Bathtub curve.

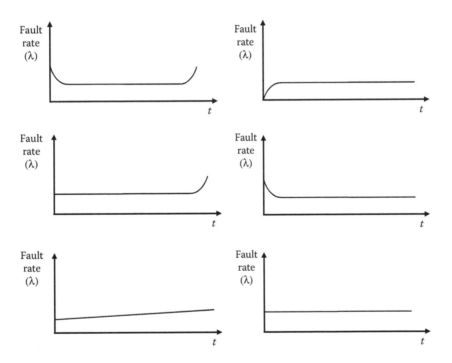

FIGURE 15.2
Bathtub and other failure rate curves.

The traditional way to look at the reliability of an equipment is through its failure rate. The most traditional approach is symbolically represented by the so-called "bathtub curve" (Figure 15.1). However, this approach has evolved into some other configurations. Moubray (1997) analyzes this subject in detail.

Figure 15.2 presents the bathtub and other failure curves.

In the following sections, a few aspects concerning quantitative reliability analysis will be summarized.

15.3 Reliability Analysis

Some of the more usual parameters related to reliability are the mean time between failures (MTBF), the mean time to repair (MTTR), and availability (A), which are calculated as follows (Ferreira, 1998):

$$MTBF = \frac{\sum_{i=0}^{n} TBF_i}{n} \tag{15.1}$$

$$MTTR = \frac{\sum_{i=0}^{n} TTR_i}{n} \tag{15.2}$$

where:
  TBF—Time between failures
  TTR—Time to repair

The operational availability is calculated by the following formula:

$$A = \frac{MTBF}{MTBF + MTTR} \tag{15.3}$$

Other parameters intrinsically linked to the preceding are the failure rate λ and the rate of repairs μ.

The failure rate λ is given by:

$$\lambda = \frac{1}{MTBF} \tag{15.4}$$

The rate of repairs μ is given by:

$$\mu = \frac{1}{MTTR} \tag{15.5}$$

The preceding parameters are determined through average values, assuming some stability in the equipment life cycle.

Another parameter that must also be taken into consideration is the Mean Waiting Time (MWT), which corresponds to the average waiting time between the fault identification and the start of the corrective maintenance intervention. The quantification of this time is important because it allows one to distinguish between the intervention intrinsic time and the total time required by the intervention.

The accompanying of the above parameters is made from their instantaneous values, as summarized below.

  F(t)—The probability density failure function
  F(t)—The accumulated failure function
  R(t)—The reliability function

The probability density failure function represents the probability the equipment will fail at time *t*. Therefore, it represents the function of instantaneous probability or the function

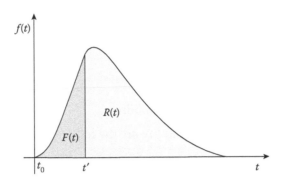

FIGURE 15.3
Probability density failure function.

of inoperability, which represents the quantity of equipment that is failing at a given time t per unit of time in relation to the initial population, not conditioned by the number of survivors in the instant before t.

For a time interval dt, it yields:

$$Prob(t \leq T \leq t+dt) = f(t) \cdot dt \tag{15.6}$$

$f(t)$ represents the probability of failure in the time interval dt.

Figure 15.3 illustrates the probability density failure function, as well as the graphical interpretation of the cumulative probability of the failure and reliability functions, which will be analyzed next.

Considering the random variable T, the time until the fault, the distribution function for this variable is defined by:

$$F(t) = Prob(T < t) = \int_0^t f(t)dt \tag{15.7}$$

The reliability, the complementary function of $F(t)$, is given by:

$$R(t) = Prob(T \geq t) \tag{15.8}$$

As a consequence, the probability density $f(t)$ is the derivative of the function $F(t)$:

$$f(t) = \frac{dF(t)}{dt} = -\frac{dR(t)}{dt} \tag{15.9}$$

and

$$F(t) + R(t) = 1 \tag{15.10}$$

Defining the instantaneous failure rate $\lambda(t)$ as the probability of the existence of failure at time t, provided there has not been any failure up to that instant, it becomes:

$$\lambda(t)dt = \frac{F(t+dt) - F(t)}{R(t)} \tag{15.11}$$

which can also be given by:

$$\lambda(t) = \frac{f(t)}{R(t)} \tag{15.12}$$

15.3.1 Statistical Methods Applied to Reliability

The most usual continuous laws in reliability are the following (Billinton and Allan, 1992):

- Negative exponential law
  The law with a parameter—λ (the failure rate)
- Logarithmic normal law
  The law with two parameters—m (average) and τ (standard deviation)
- Weibull's law
  The law with three parameters—β (shape parameter), η (scale parameter), and γ (position parameter)

However, there are other laws of great relevance, such as:

- Binomial
- Poisson's
- Normal
- Gamma
- Rayleigh
- Rectangular

The adoption of a specific statistical law is done after verifying the validity of this law through a suitability test, admitting a risk error α, which represents the level of significance. The most commonly used suitability tests are as follows:

- Test of χ^2
- Kolmogorov-Smirnov test

15.3.1.1 Exponential Distribution

The exponential distribution law applies when the failure rate λ is substantially constant over a given time interval. The associated parameters are as follows:

$$f(t) = \frac{dF(t)}{dt} = -\frac{d(1 - R(t))}{dt} = \lambda \cdot e^{-\lambda t} \tag{15.13}$$

Represents the probability of failure between t and $t + dt$.

$$R(t) = e^{-\int_0^t \lambda \, dt} = e^{-\lambda t} \tag{15.14}$$

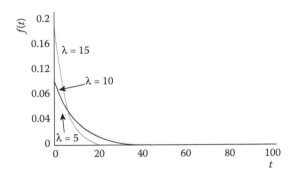

FIGURE 15.4
Probability failure function in the exponential distribution.

Corresponds to the survival probability between 0 and t.

$$F(t) = 1 - R(t) = 1 - e^{-\lambda t} = \int_0^t f(t)dt \qquad (15.15)$$

Represents the failure probability between 0 and t.

$$\lambda(t) = \frac{f(t)}{R(t)} = \frac{\lambda \cdot e^{-\lambda t}}{e^{-\lambda t}} = \lambda \qquad (15.16)$$

Represents the failure rate, which is a constant.

$$E(t) = \frac{1}{\lambda} \qquad (15.17)$$

Represents the mathematical expected value.

$$v = \frac{1}{\lambda^2} \qquad (15.18)$$

Represents the variance.

$$\sigma = \frac{1}{\lambda} \qquad (15.19)$$

Represents the standard deviation.
Figure 15.4 illustrates a probability density function for three values of the λ parameter.

15.3.1.2 Normal Logarithmic Distribution

A random variable t has a logarithmic normal distribution, with parameters μ and σ, if $\ln(t)$ has a normal distribution with parameters μ and σ. The parameters associated with this distribution are as follows:

$$f(t) = \frac{1}{t\sigma\sqrt{2\pi}} e^{-(\ln t - \mu)^2 / 2\sigma^2} \qquad (15.20)$$

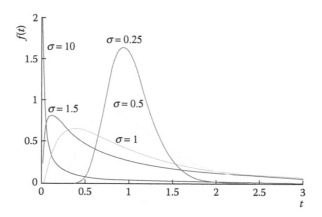

FIGURE 15.5
Probability failure function in the logarithmic distribution.

Represents the failure probability between t and $t + dt$.

$$R(t) = \int_t^\infty f(t)dt \tag{15.21}$$

Corresponds to the survival probability between 0 and t.

$$F(t) = 1 - \int_t^\infty f(t)dt \tag{15.22}$$

Gives the failure probability between 0 and t.

$$\lambda(t) = \frac{f(t)}{R(t)} \tag{15.23}$$

Represents the failure rate.

$$E(t) = e^{(\mu + (\sigma^2/2))} \tag{15.24}$$

Represents the mathematical expected value.

$$\nu = e^{(2\mu + 2\sigma^2)} - e^{(2\mu + \sigma^2)} \tag{15.25}$$

Represents the variance.

$$\sigma = \sqrt{v(t)} \tag{15.26}$$

Represents the standard deviation.
Figure 15.5 shows the probability density function for several values of σ.

15.3.1.3 Weibull's Law

Weibull's law is very flexible because, being a three-parameter law, it can be adjusted to several types of experimental and operational results.

$$f(t) = \frac{\beta}{\eta}\left(\frac{t-\gamma}{\eta}\right)^{\beta-1} e^{-((t-\gamma)/\eta)^{\beta}} \tag{15.27}$$

The preceding formula gives the probability density, where

$$t > \gamma$$

and

$$\beta > 0 \text{ form parameter (without units)}$$

$$\eta > 0 \text{ scale parameter (unit of time)}$$

$$-\infty < \gamma < +\infty \text{ position parameter (unit of time).}$$

The β parameter allows adaptation of the shape of the curves $\gamma(t)$ to the different phases of the life of a physical asset.

The scale parameter η corresponds to the time at which the probability of failure is 63.2%, indicating the number of operating units for which most of the sample elements (63.2%) will be affected.

The position parameter γ has the following characteristics:

If $\gamma > 0$, there is total survival between $t = 0$ and $t = \gamma$.
If $\gamma = 0$, the faults start at the origin of the time.
If $\gamma < 0$, the faults start before the origin of the time.

The reliability function is given by

$$R(t) = e^{-((t-\gamma)/\eta)^{\beta}} \tag{15.28}$$

And, as a consequence, the allocation function by

$$F(t) = 1 - e^{-((t-\gamma)/\eta)^{\beta}} \tag{15.29}$$

and

$$\lambda(t) = \frac{\beta}{\eta}\left(\frac{t-\gamma}{\eta}\right)^{\beta-1} \tag{15.30}$$

gives the instantaneous rate of failures, in which:

If $\beta < 1 \Rightarrow \lambda(t)$ decreases (childhood)
If $\beta = 1 \Rightarrow \lambda(t)$ constant (normal operation)
If $\beta > 1 \Rightarrow \lambda(t)$ increases (old age)

The mathematical expected value is given by

$$E(t) = \gamma + \eta \cdot \Gamma\left(1 + \frac{1}{\beta}\right) \tag{15.31}$$

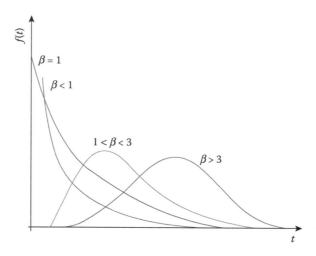

FIGURE 15.6
Probability failure function in the Weibull distribution.

The variance is given by

$$v(t) = \eta^2 \Gamma\left(1 + \frac{2}{\beta}\right) - \eta^2 \left[\Gamma\left(1 + \frac{2}{\beta}\right)\right]^2 \tag{15.32}$$

And the standard deviation by

$$\sigma = \sqrt{v(t)} \tag{15.33}$$

Figure 15.6 represents the probability density function for several values of β.

The Weibull distribution is one of the most widely used in reliability due to its great versatility, adapting to most real physical asset situations with precision. It allows characterization of faults during the phases of the life of an asset. Other distributions, such as exponential, normal, or lognormal, can be understood as particular cases of the Weibull distribution.

15.3.1.4 Serial and Parallel Systems

All equipment is implemented through serial and/or parallel systems, with the first the most relevant, as will be demonstrated next. The final equipment reliability is the result of the internal configuration, with the serial configuration the most predominant in almost all systems.

The approach presented here considers the following presuppositions:

- System reliability is evaluated at a point t in time—that is, the components present static reliabilities in t.

- The components of the system are presented in two states: operating or nonoperating.

- The components fail independently.

The characteristics of a serial system are the following:

- All components should work adequately for the system to work properly.

FIGURE 15.7
Block diagram for a serial system.

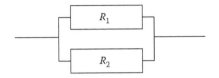

FIGURE 15.8
Block diagram for a parallel system.

The block diagram for this system is shown in Figure 15.7.
The formula that represents the functioning of a serial system is the following:

$$R(t) = R_1(t) \times R_2(t) \cdots \times R_n(t) \tag{15.34}$$

where $R(t)$ represents the static reliability at a point t in time, with $R_i(t)$ the static reliability of each component at time t.
The characteristics of a parallel system are the following:

- All components must fail for the system to fail.

The block diagram for this system is shown in Figure 15.8.
The formula that represents the functioning of a parallel system is the following:

$$F(t) = F_1(t) \times F_2(t) \cdots \times F_n(t) \tag{15.35}$$

⇔

$$1 - R(t) = (1 - R_1(t)) \times (1 - R_2(t)) \cdots \times (1 - R_n(t)) \tag{15.36}$$

⇔

$$R(t) = 1 - ((1 - R_1(t)) \times (1 - R_2(t)) \cdots \times (1 - R_n(t)) \tag{15.37}$$

A parallel system fails if all components fail, which means that the evaluation of the probability of failure is the result of the failure of all components, as represented in Equation 15.35.
According to Equation 15.10, $F(t) + R(t) = 1$. The conjugation of this equation with Equation 15.35 results in Equation 15.37, which permits evaluation of the reliability of a parallel system.

15.4 Failure Modes and Effects Analysis/Failure Modes, Effects, and Criticality Analysis

15.4.1 Background

Failure modes and effects analysis is a widespread method in the context of quality assurance and reliability of equipment and processes, from design to operation, that allows systematic evaluation of their failure modes.

This tool helps to identify faults in equipment or processes by recognizing potential failures and their signs, preferably in the phase prior to production in order to maximize their reliability.

Although the methodology was developed aimed at the design of new products and processes, FMEA, because of its great versatility, can be applied much more widely. From this perspective, it is also used to decrease the probability of failure in equipment and also in administrative processes.

FME(C)A or FMECA is an extension of the preceding concept, which, in addition to FMEA, includes a criticality analysis that is used to evaluate and sequence the probability of failure modes according to the severity of their consequences (QS 9000, 1997; Stamatis, 2003; Seet-Larsson, 2010).

FMEA and FMECA are very versatile tools that can be used for the following purposes:

- Contribute to improving the projects of equipment and processes, allowing higher reliability, quality, security, cost reduction, and increased customer satisfaction
- Contribute to optimizing maintenance plans in repairable systems and quality assurance procedures
- Contribute to the development of a knowledge base of failure modes and corrective actions to be implemented in future interventions

FMEA and FMECA can be implemented either by using a form on paper or in a spreadsheet. However, to improve the consistency of FMECA, it is necessary to build and update the equipment's historical database and prepare several reports. One of these is the common failure modes, which can be indexed to the risk priority number (RPN)—which will be discussed below—and the equipment or process performance after the implementation of corrective actions.

There are several guidelines and standards published for the requirements and forms of FMEA and FMECA, as referred to in Chapter 17, Section 17.4.

15.4.2 Failure Modes and Effects Analysis

Usually, FMEA is supported by a group of people (the FMEA working group) to identify, for equipment or a process, its functions, the types of failures that can occur, their effects, and their possible causes. Then, the risks for each cause of failure are evaluated through indexes and, based on this assessment, the necessary actions to reduce them are found in order to increase the equipment or process reliability.

FMEA can be subdivided into two groups:

1. FMEA of equipment—In this group, the failures that may occur with the equipment under study are considered. The purpose of this analysis is to eliminate the failures in the project phase. This analysis is usually designated as FMEA of the project.
2. FMEA of a process—In this group, the faults inherent to the planning and execution of the process are considered. The purpose of this analysis is to avoid the faults of the process, with the objective of eliminating nonconformities of the equipment with the project specifications.

FMEA is very versatile and may, in addition, reduce the probability of damage due to new projects of equipment or processes. It also reduces the probability of failures in equipment and processes already in operation, taking into account the analysis of the historical failures that have occurred.

FMEA can also be applied to administrative procedures that, despite being a less common situation, allow examination of the potential failure of each step of those with the same purpose as the previous situations, that is, to reduce the risk of failure.

The principle of this methodology is the same, regardless of the type of application of FMEA, that is, FMEA of equipment, process, or procedure, applied either to new products or processes or those already in operation.

The stages of implementation for FMEA are (Toledo and Amaral, 2001):

1. Planning—This phase includes the following steps:
   - Description of the objectives and definition of the scope of the analysis— Includes the definition of equipment or processes to be analyzed
   - Formation of the working group—Includes the definition of the members who will join the group, which should be small (four to six people) and multidisciplinary (must integrate people from different areas such as quality, development, and production)
   - Planning meetings—Includes timely scheduling of the meetings in order that all members will be present
   - Preparation of the documentation—Includes collecting the necessary information for FMEA

2. Analysis of the potential failures—This phase is carried out by the working group that analyzes the problems and meets the FMEA forms according to the following steps:
   - Identification of the functions and features of the equipment or process
   - Characterization of the types of potential failures for each function
   - Identification of the effects for each type of failure
   - Identification of possible causes for each failure
   - Characterization of the current fault controls

3. Risk assessment—This phase defines three types of indices, severity (S), occurrence (O), and detection (D), for each cause of failure, according to the predefined criteria (Tables 15.1 through 15.3 illustrate an example of criteria that can be used for the preceding indices, with a four-point scale). Next, the risk priority numbers are calculated by multiplying the previous three indices (RPN = S × O × D). For the evaluation of these indices, the following aspects should be taken into consideration:
   - The evaluation of each index is independent; that is, when the working group is examining an index, the others cannot be taken into account—for example, when evaluating the severity index of a particular cause of a fault of equipment whose effect is significant, this index cannot be assigned a lower value to induce a higher probability of detection.
   - In the case of FMEA of process, the equipment's capability index C_{pk} may be used to determine the occurrence rate, taking into account that this index indirectly measures the adjustment and dispersion relative to the average process. This dispersion is the one that is usually used in the control charts: $3\sigma - 3$ times the standard deviation above and below the average value—(C_{pk} is a measurement that relates the actual yield of a piece of equipment or a process to its specified yield under quality control. It is expected as a minimum requirement that the product specifications be contained inside the limits $\pm 3 \cdot \sigma$, with $C_{pk} = $ (Upper control limit − Lower control limit)$/6 \cdot \sigma$).

TABLE 15.1

Examples of Severity Indices

Severity		
Index	Severity	Criterion
1	Minimum	The customer almost does not notice the occurrence of failure.
2	Slight	Small deterioration in the performance identified by the client, with little dissatisfaction for him or her.
3	Moderate	Significant deterioration in performance, with some customer dissatisfaction.
4	High	Total dysfunction, manifesting as great customer dissatisfaction, and may affect security.

TABLE 15.2

Example of Occurrence Rates

Occurrence		
Index	Occurrence	C_{pk}
1	Remote	$C_{pk} > 1{,}33$
2	Small	$1{,}33 \geq C_{pk} > 1{,}00$
3	Moderate	$C_{pk} \leq 1{,}00$
4	High	$C_{pk} \leq 1{,}00$

TABLE 15.3

Example of Detection Rates

Detection		
Index	Detection	Criterion
1	Very high	Very high probability of being detected.
2	High	High probability of being detected.
3	Moderate	There is some probability of being detected.
4	Small	There is a low probability of being detected.

4. Improvement—In this phase, the working group, using its knowledge, creativity, and techniques, such as brainstorming, lists the actions that can be implemented with the aim:

- To prevent failures
- To prevent the causes of failures
- To inhibit the occurrences of failures
- To limit the effects of failures
- To increase the probability of detection of the causes of failures

These actions shall be reviewed to evaluate their viability with the aim of deciding which ought to be implemented. One way to control the result of the application of these measures may be through the FMEA form, in which columns can be used to record the measures recommended by the working group, the name of the person responsible for the implementation of each, its execution time, and, consequently, those that will have a new reassessment.

5. Monitoring—The FMEA form shall be a dynamic document; that is, once an analysis of an equipment or process is performed, it must be reviewed whenever changes occur in it. Beyond this, even if there are no changes, a regular review and analysis shall be done, comparing the potential failures with those that actually occur in reality in order to allow a comparative analysis or the incorporation of unexpected failures.

6. Relevance—The FMEA methodology is important because it can provide:
   - A systematic cataloging of information about the failure of equipment or processes
   - A better understanding of the problems in equipment or processes
   - The implementation of continuous improvement actions on the equipment or process, based on evidence and properly monitored
   - Cost savings through the prevention of failures
   - The increase of quality and the image of the organization as the result of an attitude that prevents faults and increases cooperation and teamwork, as well as customer satisfaction

15.4.3 Failure Modes, Effects, and Criticality Analysis

FMECA allows, in addition to the features of FMEA, highlighting of the failure modes that have a high probability of occurrence and the severity of their consequences, allowing sequencing of their solutions according to the relevance of the occurrence probabilities.

As part of project design, the purpose of FMECA is to eliminate failure modes with a high probability of having a high severity and reduce as much as possible the ones that have a high probability of occurrence.

As mentioned before, the sequencing of priorities is done through the risk priority number, which is the result of the multiplication of the indices of severity (S), occurrence (O) and detection (D): $RPN = S \times O \times D$:

- The RPN is the mathematical product of relevance (severity) of an effects group, the probability of failure that can create causes associated with these effects (occurrence), and the ability to detect the failure before the occurrence (detection).
- The RPN is used to help identify the most important risks and implement the appropriate corrective actions. However, severity, occurrence, and detection may not have equal weight in terms of risk, with differences inherent to the nonlinear nature of the individual levels. As a result, some results of the $S \times O \times D$ product can produce RPNs lower than other combinations, but actually have a higher risk.

FMECA Analysis										
Equipment/Process:						**Location:**				
Responsible / Group:						**Date:**				
Code	Designation	Failure Modes	Effects of faults	Severity (S)	Service level	Occurrence (O)	Detection means	Detection (D)	RPN	Solutions

FIGURE 15.9
Example of an FMECA form.

FMECA Analysis										
Equipment/Process: Filament lamp						**Location:** Laboratory of Quality				
Responsible / Group: Felipe / Mikael; John; Candid; Rafael; Saints						**Date:** 13.07.2016				
Code	Designation	Failure Modes	Effects of faults	Severity (S)	Service level	Occurrence (O)	Detection means	Detection (D)	RPN	Solutions
3745	Lamp	Composition outside the specified	Illuminance levels and energy consumption	8	Quality loss	3	Luximeter	6	144	Quality control of raw materials
		Diameter outside the specified	Idem	6	Idem	4	Idem	5	120	Settings of equipment manufacturing

FIGURE 15.10
Example of filling out an FMECA form.

Figure 15.9 illustrates an example of a FMECA form for evaluation of the RPN, and Figure 15.10 shows a filled form.

High levels of severity should be given special attention, particularly when associated with high values of occurrence. To emphasize the importance of these combinations, an intermediate parameter has been defined called criticality relative to the RPN, which is defined as the mathematical product of the occurrence and severity ($C = O \times S$). However, severity and occurrence remain uneven in terms of risk, and their levels remain nonlinear.

It is usual to define the indices on a scale from 1 to 10, so, in this case, the maximum value the RPN can have is $10 \times 10 \times 10 = 1000$. This means that a fault with this score is very severe and its occurrence is almost certain. If the occurrence is very scarce, it corresponds to $O = 1$, and the RPN will decrease to 100.

OCCURRENCE	SEVERITY			
	1 - Insignificant	2 - Marginal	3 - Critical	4 - Catastrophic
4 - Frequent	4	8	12	16
3 - Likely	3	6	9	12
2 - Remote	2	4	6	8
1 - Unlikely	1	2	3	4

FIGURE 15.11
Example of an array of criticality (C).

OCCURRENCE	DETECTABILITY															
	1	2	3	4	1	2	3	4	1	2	3	4	1	2	3	4
4 - Frequent	4	8	12	16	8	16	24	32	12	24	36	48	16	32	48	64
3 - Likely	3	6	9	12	6	12	18	24	9	18	27	36	12	24	36	48
2 - Remote	2	4	6	8	4	8	12	16	6	12	18	24	8	16	24	32
1 - Unlikely	1	2	3	4	2	4	6	8	3	6	9	12	4	8	12	16
	1 - Insignificant				2 - Marginal				3 - Critical				4 - Catastrophic			
	SEVERIDADE															

FIGURE 15.12
Example of a RPN matrix.

In order to show an example with a different level of scale, Figures 15.11 and 15.12 illustrate the matrices of criticality and RPN, respectively, with maximum values for severity, occurrence, and detectability of 4.

In the case illustrated in the example of Figure 15.12, one can define the RPN levels as follows:

- Intolerable—The cause of the problem must be eliminated (RPN $\geq$ 32).
- Undesirable—This should only be accepted when reducing the potential risk is impractical and with the consent of those responsible ($24 \leq$ RPN < 32).
- Tolerable—Acceptable with adequate control and the agreement of those responsible ($12 \leq$ RPN < 24).
- Permissible (RPN < 12).

As can be seen, FMEA and FMECA correspond to extremely powerful techniques that allow solving complex problems in a simple way. The identification of the main causes of the faults as well as their sequencing may be the first step in its resolution. However, in more complex situations, it may be necessary to use other types of reliability analysis, like dynamic modeling.

15.5 Reliability, Availability, Maintainability, Safety

This section presents a short summary of RAMS analysis, whose importance corresponds to the optimal combination of these four areas, aiming toward rationality of development costs, performance, and operation throughout the life cycle of physical assets.

According to NP EN 50126:2000, "RAMS can be characterized as a qualitative and quantitative indicator of the degree of reliability in which the system or subsystems and components that compose the system can operate as required, being available and safe."

The RAMS concept emphasizes the variables reliability, availability, maintainability, and safety in the analysis of the performance of equipment or systems, considering their implications at the technical, social, and economic levels. At the same time, it enables the

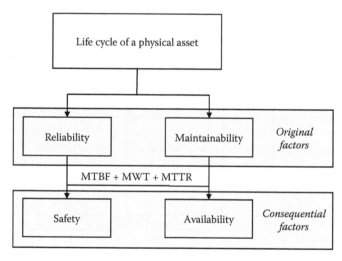

FIGURE 15.13
Relationships among variables R, A, M, and S.

RAMS variables	Implications
Reliability	• Availability • Safety • Quality of maintenance interventions • LCC (Life Cycle Cost)
Availability	• Compliance with the functions of the Asset according to specified • LCC
Maintainability	• Availability • Duration and cost of interventions • LCC
Safety	• Protection of users • Environmental protection

FIGURE 15.14
R, A, M, and S variables and their implications on assets.

establishment of metrics to evaluate the performance of equipment and systems through the use of ratios that characterize each variable.

Figures 15.13 and 15.14 illustrate the relationships among those variables and their implications. On this subject, see, for example, Villemeur (1992) and Stapelberg (2009), and also Section 17.5 of Chapter 17 of this book.

The evaluation conducted for all four variables of RAMS analysis is complex and requires a specific approach, which, by extension, is outside the scope of this book.

15.6 A Case Study

The example presented here refers to an electrical-source energy power system in a hospital in Europe. The system has several levels of redundancy and is controlled by an intelligent load switch, as is shown in Figure 15.15.

The system has the following main components:

- An external power source
- Two power switches and two transformers in one cell
- One power switch and one transformer in one cell
- Three generators powered by diesel engines
- Three boards with switching equipment
- An intelligent controller

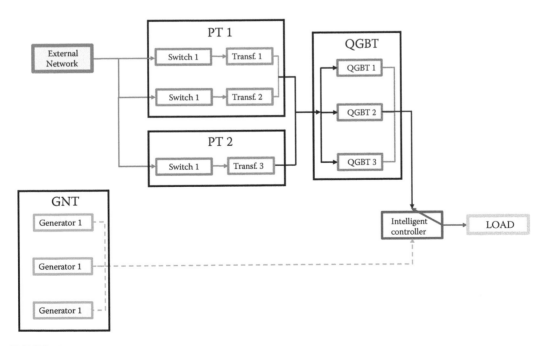

FIGURE 15.15
Series-parallel power system.

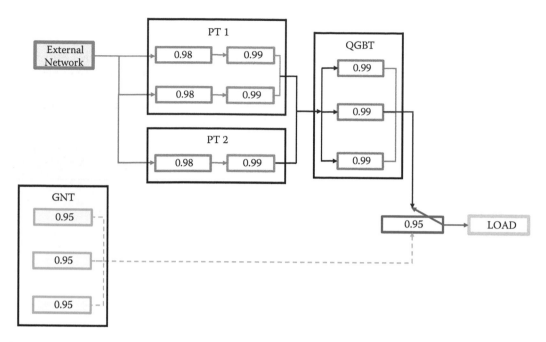

FIGURE 15.16
Reliability values of a series-parallel power system.

The question is to evaluate the general reliability of the system, considering all components in its useful life, when all fault rates are constant, as shown in Figure 15.16.

Appling Equations 15.34 (serial systems) and 15.37 (parallel systems), the global reliability for the electrical energy source power system can be found.

Component GNT:

$$R_{GNT} = 1 - ((1 - 0.95) \times (1 - 0.95) \times (1 - 0.95)) = 0.99989$$

Component PT1:

$$R_a = 0.98 \times 0.99 = 0.97020$$

$$R_b = 0.98 \times 0.99 = 0.97020$$

$$R_{PT1} = 1 - ((1 - 0.97020) \times (1 - 0.97020)) = 0.99911$$

Component PT2:

$$R_{PT2} = 0.98 \times 0.99 = 0.97020$$

Component QGBT:

$$R_{QGBT} = 1 - ((1 - 0.99) \times (1 - 0.99) \times (1 - 0.99)) = 0.99999$$

Components PT1//PT2:

$$R_{PT1//PT2} = 1 - ((1 - 0.99911) \times (1 - 0.97020)) = 0.99997$$

Components PT1//PT2-QGBT:

$$R_{PT1//PT2\text{-}QGBT} = 0.99997 \times 0.99999 = 0.99996$$

Components PT1//PT2-QGBT-intelligent controller(IC)-GNT:

$$R_{PT1//PT2\text{-}QGBT\text{-}IC\text{-}QNT} = 0.99996 \times 0.95000 \times 0.99989 = 0.94986$$

This simple example based on a real case demonstrates how much a weaker component influences the global reliability of the system—in this case making it decrease a lot.

16

Management Methodologies

16.1 Background

With the evolution of several maintenance concepts and the development of new approaches and methodologies applied to other aspects of management, in particular in the area of quality and production, maintenance activity has added and adapted many new concepts such as Lean maintenance, 5S, the PDCA cycle, and SWOT analysis.

Whatever the perspective from which they are seen, in practice they represent important contributions to maintenance quality improvement.

16.2 5S

A mandatory methodology for the organization of any plant is a system of Japanese origin called 5S because it is based on five principles or senses beginning with the letter S (Osada, 1991) that mean:

1. Seiri—Sense of use—Refers to the practice of checking all tools, materials, and so on, on the desktop, keeping only the essentials for the tasks that are being performed. Everything else is stored or discarded. This process leads to a reduction of barriers to labor productivity.

2. Seiton—Sense of order—Emphasizes the need for an organized workspace; that is, the provision of tools and equipment must comply with order to allow an easy flow for development of tasks. Tools and equipment should be left in places where they will be used later. The process should be done in order to eliminate unnecessary movements.

3. Seiso—Sense of cleaning—Describes the need to keep the workspace as clean as possible. Cleaning in Japanese companies is a daily activity. At the end of each working day, the working space is cleaned and everything is put back in its right place, making it easy to know what corresponds to each site and what is essential. The importance of this procedure is to remember that cleaning should be an integral part of daily work, not a mere casual task when objects are disorderly.

4. Seiketsu—Sense of health—Refers to the standardization of work practices, such as keeping similar objects in similar locations. This procedure leads to a work practice,

standard layout, and practices favorable to physical, mental, and environmental health;

5. Shitsuke—Sense of self-discipline—Refers to maintaining and reviewing standards. Since the previous 4Ss have been established, they become a new way of working and should not allow the return to old practices. However, when there is a new improvement or tool, or the decision to implement new practices, it is advisable to review the previous four principles.

The objective of the 5S methodology is to improve efficiency through the proper definition of the objectives in the use of materials, identifying the unnecessary ones and stressing the importance of organizing, cleaning, and identifying materials and workspaces as well as the maintenance and also the improvement of the 5Ss themselves. The main benefits of this approach are the following:

- Increased productivity by reducing wasted time looking for objects—The necessary objects must be at hand
- Reduced costs and better use of materials—Excessive accumulation of materials provides degeneration and unnecessary costs
- Improving the quality of products and services
- Fewer work accidents
- Increased satisfaction of people with their work

The next sections present a detailed description of 5S as well as the added value that each one can bring to the organization.

16.2.1 First S—Seiri—Sense of Use

The sense of use can also be interpreted as storage and organization. With the implementation of the first S, the space and working methods start to be put in order to use only what is really necessary and applicable. Therefore, it is essential to have only what is necessary, in the proper amounts and procedures, in order to optimize interventions.

It is critical to separate and classify useful from useless objects, as follows:

- What is always used—Place near the workplace
- What is almost always used—Place near the workplace
- What is occasionally used—Place a little away from the workplace
- What is rarely used but necessary—Place separately in a given location
- What is unnecessary—Should be removed because it takes up space needed for other objects and difficult work

The advantages of the implementation of the sense of use are the following:

- Reduces the need for space and costs, diminishes stock, and optimizes transportation, among others
- Facilitates physical organization, production control, and optimization of maintenance activity

- Optimizes the purchase of materials and waste in stored materials
- Increases the productivity of the equipment and the people involved
- Increases the sense of humanization, organization, and economy
- Reduces physical fatigue and facilitates operations
- Decreases the probability of accidental risks

16.2.2 Second S—Seiton—Sense of Order

The sense of order can also be defined as systematization, classification, and cleaning. The goal is to identify and arrange everything so anyone can easily find what they need and easily visually identify each object.

For the implementation of the second S, it is important to:

- Standardize the nomenclatures
- Use labels and bright colors to identify objects, following a pattern
- Store different objects in different places
- Visually indicate the critical points, such as fire extinguishers, high-voltage sites, parts of equipment that involve risks, and so on
- Determine the storage location of each object
- Never leave objects or furniture in passages, interrupting the movement of people and equipment

The advantages of the implementation of the sense of order are the following:

- Reduce search time to make interventions, operate tools, and so on
- Reduce need for stock control and production
- Facilitate internal logistics, document control, and management of files, facilitating the execution of the work within the planned time frame
- Prevent unnecessary purchase of components and damage to stored items
- Allow greater rationalization of work, less physical and mental fatigue, and a better environment
- Allow a better arrangement of furniture and equipment
- Facilitate workplace cleaning

The efficient ordering of objects necessary to execute the work should be implemented with a standardized nomenclature and disseminated to all stakeholders in the appropriate places, such as documents, folders, files, meeting rooms, and so on, with the correct indication of where they belong. People must know where to look for each object when necessary, and everyone must follow the same rules.

16.2.3 Third S—Seiso—Sense of Cleaning

The sense of cleaning means that every person shall be aware of the importance of living in a clean environment and the benefits resulting therefrom. A clean environment means quality and safety.

The advantages of the implementation of the cleaning sense are the following:

- Increase people's productivity, equipment, and materials
- Minimize losses and object damage—It is important that people be aware of them and adopt new habits, such as:
  - Clean the equipment after each use, and leave it in the appropriate operating conditions in order for the next user to find it clean
  - Learn not to create a messy environment, and eliminate the causes of dirt
  - Define those responsible for each area and their functions
  - Keep equipment and tools in the best condition possible
  - Keep the workplace clean, including locations such as the corners and ceilings
  - Do not throw trash on the floor
  - Put trash in the appropriate places, according to its nature

Under the ambit of the third S, it shall also include the maintenance of data and updated information, and to be honest in the workplace, including having a good relationship with colleagues.

16.2.4 Fourth S—Seiketsu—Sense of Health

The sense of health also includes hygiene—it means the maintenance of cleanliness and order in the workplace. Each aspect must be good itself, and this also means a way of life, which underlies the increase of quality of products and services.

People should be aware of the importance of the implementation of this phase, in particular by adopting a set of measures such as the following:

- To have the three other Ss previously implemented
- To train people to learn and apply knowledge to assess whether the concepts are correctly applied
- To eliminate unsafe working conditions in order to avoid accidents or dangerous manipulation
- To humanize the workplace through a harmonious and humane coexistence
- To respect colleagues as people and professionals
- To constructively collaborate with co-workers
- To comply with schedules
- To deliver documents and materials on time

The benefits of the implementation of sense of health are the following:

- Better security and performance of staff
- Better prevention of damage to the health of people in the workplace
- Better image of the organization, both internally and externally
- Better level of staff satisfaction and motivation

At this stage, warnings and instructions shall also be placed to prevent errors in work operations. Warnings should be visible from a distance, well featured, and accessible to all people. It is also important to check that the program of 5S is actually being implemented, verifying each step, and whether people are prepared and motivated to carry out the program.

16.2.5 Fifth S—Shitsuke—Sense of Self-Discipline

The sense of self-discipline means:

- Use constructive creativity at work
- Improve communication among people in the workplace
- Share the vision, mission, and values of the organization, harmonizing goals
- Train staff, with patience and persistence, raising awareness among them for the implementation of 5S
- Regularly evaluate the implementation of 5S to correct any deviations

It is important to follow the operating procedures and ethical standards of the organization, always looking for continuous improvement. Self-discipline implies an attitude of continuous improvement for all. Awareness of the importance of quality excellence is the key.

The advantages of the implementation of the sense of self-discipline are the following:

- Reduce the need for constant monitoring
- Facilitate the implementation of all operations
- Prevent the loss of efficiency of work and time
- Bring predictability to the outcome of operations

Figure 16.1 illustrates the relationships among the 5Ss.

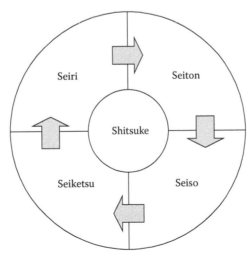

FIGURE 16.1
Relations among the 5Ss.

16.3 Lean Maintenance

Lean thinking is a way to specify value, line up the best sequence of actions that create value, perform activities without interruption every time someone requests, and perform them more effectively.

Lean maintenance (LM) can be defined as the proactive realization of the planned activities, that is, using strategies developed by the appropriate application of TPM, RCM, 5S, analysis of failures (FMECA), condition-based maintenance (predictive), and CMMS (Heisler, 2003; Levitt, 2008).

Lean maintenance underlies the concept of Lean thinking, whose main implementation tools are the following:

    i. Value stream mapping (VSM)
   ii. Lean metrics
  iii. Kaizen
   iv. Kanban
    v. Standardization
   vi. 5S
  vii. Setup time reduction
 viii. TPM
   ix. Visual management
    x. Poka-yoke (mistake proofing)

i. *Value Stream Mapping*
VSM corresponds to a set of actions that add and do not add value, but are required to enable the product:

- From design to product launch
- From the request to delivery
- From raw material to the consumer

It is composed of:

- Information flow
- Material flow

VSM is a tool that helps identify the flow of material and information within an organization. It should follow the production flow of a product, from the consumer to the supplier, and carefully draw up a visual representation of every process in the material and information flow. Sometimes, it is necessary to detail a specific process for a specific purpose to reduce waste disposal.

The control of the activities of a manual operator needs to be carefully analyzed, because it is where the most waste can be identified. It is necessary to analyze each segment in trying to reduce waste and reduce operating time and main costs. These improvements must be translated to the final consumer.

ii. *Lean Metrics*

There are several ways to evaluate Lean performance. The following ones correspond to a possible approach:

- Increased cash flow, measured through sales growth, sales by person, and daily production per hour
- Increased sales and market share, measured by days of inventory, delivery on time, and the material in process
- Continuous optimization culture, measured by delivery on time, door-to-door time, and the first through in the cell/process
- Measure of customer satisfaction, the first through, and the equipment effectiveness in the process
- Measure of sales per employee and the average cost per unit
- Measure of the days above the term of the receivable accounts

iii. *Kaizen*

Kaizen is an "umbrella" that covers all the improvement techniques, coalescing them harmoniously to make the most of what each one offers.

The 10 principles of Kaizen are the following:

1. Emphasizes the customer
2. Promotes continuous improvement
3. Openly recognizes problems
4. Promotes opening
5. Creates work teams
6. Manages projects through cross-functional teams
7. Promotes an adequate relationship process
8. Develops self-discipline
9. Informs all employees
10. Empowers all employees

The specific features of Kaizen are the following:

- Involvement of employees through suggestions
- Application of social intelligence in working time
- Management by processes
- Use of easy techniques (quality tools)
- Attack problems at the root, identifying their causes
- Build product quality to satisfy customers

The Kaizen methodology is the following:

- Define the problem—Identification of opportunities for improvement. Project selection and team training
- Measurement—Process mapping, selection of indicators, and data collection
- Analysis of current process—Identification and prioritization of the causes of the problem
- Improvement—Generation, selection, and implementation of improvement measures
- Control—Evaluation of the results and process documentation (standardization). Monitoring to ensure the maintenance and development of improvements

iv. *Kanban*
Kan = card; ban = signal.

A card system controls production and inventory, and a visual system pulls the system, as opposed to a black box that pushes the system (i.e., Material Requirements Planning [MRP]).

The Kanban system works based on using signals to enable the production and movement of items in the factory:

- The signs are conventionally made based on Kanban cards and Kanban door panels. Other means than cards may be used to pass this information
- Conventional Kanban cards are made from durable material to withstand handling due to the constant movement between customer and stock supplier, or, nowadays, electronic panels are used (eKanban)
- Each company prepares its own cards to deploy its Kanban system according to its specificities

v. *Standardization*
Standardized refers primarily to the work force, based on three elements:

1. Takt Time—It is the rate at which products must be made in a process to meet industrial requirements
2. Precise Work Sequence—Each operator performs tasks within the Takt Time
3. Standard Inventory—This includes the equipment required to keep the process operating efficiently

vi. *5S*
The 5Ss were analyzed in detail in Section 16.2.

vii. *Setup time reduction*
Setup time reduction is one of the principles of the single-minute exchange of die (SMED) system to reduce or eliminate changeover time.

Some benefits of setup time reduction are the following:

- Respond to customer needs with more flexibility
- On-time delivery

- Decrease costs due to excess inventory
- Increase production line and equipment capacity levels
- Increase changeover accuracy
- Reduce startup defects

viii. *Total Productive Maintenance*
The TPM concept was analyzed in detail in Chapter 2.

ix. *Visual Management*
The purpose of visual management is to improve the effectiveness of communication and reaction inside an organization.

Visual aids can convey messages more quickly and draw more interest from people than written information.

The main objectives of visual management are the following:

- Clearly expose waste
- The work area communicates in simple terms
- The goals are clearly indicated
- The status of production ought to be aligned with the goals
- Simple problems have solving tools on display
- Increases communication inside the organization

x. *Poka-Yoke (Mistake Proofing)*
Poka-yoke is a failsafe system to prevent the occurrence of defects in manufacturing processes and/or in the use of products. This concept is part of the Toyota production system (TPS) first developed by Shigeo Shingo.

There are two ways how poka-yoke can be used to correct errors (Shingo, 1996):

1. Control method—When poka-yoke is activated, the equipment or processing line stops, so that the problem must be fixed.
2. Warning method—When poka-yoke is activated, an alarm sounds or a light signals, aiming to alert the operator.

Poka-yoke itself is not a system of inspection, but a method for detecting defects or errors that can be used to satisfy a particular function.

The first step in the selection and adoption of effective quality control methods is to identify the inspection system that best meets the needs of a particular process.

It is based on the preceding tools that Lean maintenance can be implemented. From this broad perspective, it appears that the original Lean concept includes a lot of management tools and concepts.

Lean maintenance contributes decisively to the achievement of the objectives of production, that is, of its working methods, which must be conducted with determination and rigor, that ensure assets and production processes are in line with the indicators of availability, reliability, and productivity to ensure the overall effectiveness of the assets.

Another important feature of Lean maintenance is the constant search for the best practices in maintenance teams, aiming for the optimization of their work processes, using properly documented procedures and optimizing human and material resources and outsourcing while maintaining appropriate training of employees, always looking for the best KPI.

Recently, Lean maintenance has been one way to search for and eliminate waste and add value. It is a management concept to simplify the way materials and information are managed. The seven wastes usually considered are the following:

1. Overproduction
   • Overproduction is the largest source of waste.
2. Inventory
   • The reduction of stock occurs by identifying its root cause.
3. Transport
   • Transport never generates aggregated value to the product.
4. Standby time
   • Standby time refers to materials that are waiting in queue to be processed.
5. Movement (people)
   • Travelling to the office, copier, or store.
6. Processing excess
   • Activity or effort that is not requested by the client and does not add value to the product or service, such as performing operations resulting from needless projects and processes.
7. Defects (error correction)
   • Producing defective products means waste materials, hand labor, handling of defective materials, and so on.

Lean maintenance aims to contribute to achieving the objectives of production; that is, it seeks to align assets and productive work processes with availability, reliability, and productivity indicators, ensuring the overall effectiveness of assets.

16.4 A3 Method

The A3 method (also called A3 report) was born at Toyota and consists of a framework of A3-size paper sheets on which the problem to be solved or project to be executed is described, with its analysis, corrective actions, and action plans, using graphics and images wherever possible. It also allows documentation of the results of the efforts to solve the problems concisely as well as the methodology used for them, which involves a deep knowledge of how the work is done (Sobek and Smalley, 2008).

When implemented properly, this method allows the organization to support a systemic vision, holistic instead of punctual, because the person responsible for solving the problems seeks the consensus of all parties affected by it.

With this method of general application, the whole organization can use it in all departments, including the maintenance department.

Figure 16.2 illustrates the application of the A3 method according to Toyota.

Subject:	Team:			
History:	**Objectives:**			
• Fault History • Fault Importance • Situation context	• Scheme with the objectives to be achieved (time, interventions, costs) • Alternative Plans			
Current situation:	**Plan to implement:**			
• Diagram of the current situation • Explain the problem • What, where, when, how, how much?	What? (Interventions to be performed)	Who? (Who is responsible?)	When? (Schedule)	Where? (Location)
Cause-Effect Analysis:	**Following:**			
• What is the effect? • What is the cause? • Because? (Ishikawa Diagram)	Plan: • How to check the results? • When to check the results?	Current results: • Meeting deadlines? • Results versus expectations?		

FIGURE 16.2
Scheme of A3 method according to Toyota.

16.5 Gravity, Urgency, Tendency Matrix

The GUT matrix is another tool, not as widespread as the previous ones, but with applications in many situations because of its simplicity—it can also be used in association with FMECA.

The GUT matrix is a tool for quickly sequencing problems, taking into account their severity, urgency, and tendency:

- Gravity (G)—Defines the impact of the problem on things, people, results, processes, or organizations and the effects that may arise in the long term if the problem is not resolved
- Urgency (U)—Sets the available or necessary time to solve the problem
- Trend (T)—Defines the potential evolution of the problem, assessing growth trends, and the possible reduction or disappearance of the problem

The GUT matrix usually uses a score of 1–5 points for each dimension of the array to allow sorting the points of the problems to be treated to solve the situation in descending order (Table 16.1).

This type of analysis should be done by a working group consisting of technicians involved in the situation so that, in a pragmatic way, they can create the sequencing of the problems—there must be a consensus among the group members. Figure 16.3 presents an example of a framework to support the implementation of the GUT matrix.

TABLE 16.1
Calculation Table for GUT Matrix

Points	Severity	Urgency	Trend
5	Losses or consequences are extremely serious.	Immediate action is needed.	The deterioration will be immediate.
4	Losses or consequences are very serious.	The intervention has some urgency.	The deterioration will increase in the short term.
3	Losses or consequences are serious.	The intervention ought to be performed as soon as possible.	The deterioration will increase in the medium term.
2	Losses or consequences are minor.	Intervention may wait a bit.	The deterioration will worsen in the long term.
1	Losses or consequences are not serious.	The intervention is not urgent.	The deterioration will not increase and even may improve.

Company:						
Process / System / Equipment:						
	Problems	G	U	T	Total	Sequence
1						
2						
3						
4						
5						
6						
7						

FIGURE 16.3
Framework to support the implementation of GUT matrix.

After assigning the score, the calculation of $G \times U \times T$ must be performed and, according to the result, the definition of the sequencing.

16.6 Six Sigma

The statistical model control called 6σ (6 Sigma) was created by the Japanese quality manager Joseph Juran in the mid-1980s, having been used by large multinationals such as Motorola, 3M, Apple Computers, HP, Vodafone, and DuPont, as well as governmental organizations such as the US Army and NASA (Juran, 1992).

Six Sigma is an innovative methodology focused on eliminating defects caused within the processes in an organization, aiming to provide products or services close to perfection. Its importance in maintenance activity relates to condition monitoring by relating the

FIGURE 16.4
Six Sigma cycle.

conformity of the products and the equipment condition. It can also be used in FMECA and in particular in the case of the FMEA of processes.

The term "6 Sigma" comes from the normal distribution, representing the desired change in the form of processes to ensure the desired product quality or customer service. Quality control charts are paradigmatic cases of the concept of 6 Sigma, that is, three standard deviations above and three standard deviations below the mean ($\mu \pm 3\sigma$).

The methodology 6σ goes through the following stages (Figure 16.4):

i. Definition—Identification of problems and processes

ii. Measurement—Current characterization of the process

iii. Analysis—Study of the impact of each variable on the process

iv. Improvement—Performing simulations using mathematical models

v. Control—Monitoring the improvement process

In addition to this methodology, the PDCA cycle is also decisive in the implementation of the 6 Sigma method.

General Electric (GE), led by Jack Welch, was the main driving force in the use of 6 Sigma, and during the first five years of its implementation, GE had a profit of about 10 billion dollars (Welch and Welch, 2005).

16.7 Plan, Do, Check, Act

The PDCA cycle is another interesting tool in industrial management. It is a method of analysis and improvement created by Walter Shewhart in the mid-1920s and spread throughout the world by Deming.

It is a very useful tool to analyze and improve processes and the effectiveness of teamwork and is critical in support of management and decision-making both for the maintenance department and the organization (Tapping, 2008).

The steps of the PDCA cycle are the following (Figure 16.5):

- Plan
- Do
- Check
- Act

Plan

- Define the objectives to be achieved.
- Set the method to achieve the objectives.

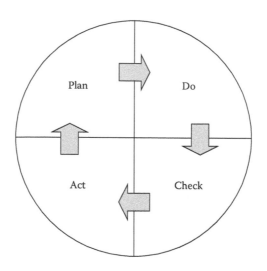

FIGURE 16.5
PDCA cycle.

Do

- Perform the tasks exactly as planned at the planning stage.
- Collect data to use in the next phase of the process.

Check

- Check if what was performed is according to plan, that is, if the objective was achieved according to plan.
- Identify deviations from the objectives.

Act

- If deviations are identified, it is necessary to define and implement actions to eliminate their causes.
- Otherwise, preventive work must be done by identifying deviations that may occur in the future as well as their causes and solutions.

The steps that are usually used to implement the PDCA methodology are the following:

   i. Identification and description of the problem
  ii. Problem understanding
 iii. Objective to achieve (Plan)
  iv. Identification of causes
   v. Tasks to be done (Do)
  vi. Characterization of the results (Check)
 vii. Standardization and training of team members for the new methodology (Act)
viii. Success recognition and sharing

16.8 Ishikawa Diagram

The Ishikawa diagram, also known as a "cause and effect diagram" or "fishbone," is a graphical tool used to support decision makers, both in management and quality control and both in production and maintenance. This diagram was proposed by the chemical engineer Kaoru Ishikawa in 1943 and improved in the next years (Kume, 1985).

This diagram is also known as 6M, because, in its initial structure, all kinds of problems could be classified into six types, as follows (Figure 16.6):

   i. Methods

   ii. Materials

   iii. Manpower

   iv. Machinery

   v. Measurements

   vi. Mother Nature

This tool allows one to hierarchically structure the potential causes of a particular problem or opportunity for improvement, as well as its effects on the quality of products or services.

Kaoru Ishikawa also noted that, although not all problems can be solved by this tool, at least 95% could be. But, any worker with few academic skills can use it.

The implementation of the Ishikawa diagram has no limits: organizations can identify, adapt, and demonstrate in specific diagrams the origin of each one of the causes of the effect and the causes that preceded those causes of effect, to a level of detail they consider appropriate. The depth of detail can be crucial to reach a better quality of results and their analysis. The more information available about the problems and their causes, the better the chances of solving them.

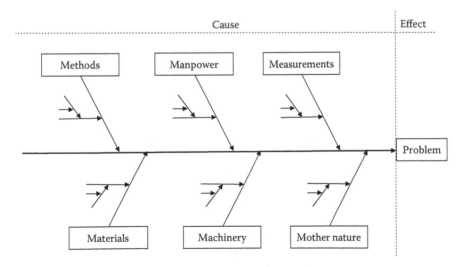

FIGURE 16.6
Ishikawa diagram.

16.9 Brainstorming

Brainstorming is a tool to explore the creative potential of an individual or a team, with a view to achieving predetermined objectives.

The brainstorming method was created by Alex Osborn, originally in the United States, in the area of human relations and advertising, among other fields of activity, with maintenance a good example (Osborn, 1963).

The brainstorming method proposes that a group of people, from 2 to 10, meet and use the differences between their ideas, with the objective of reaching a common denominator that is effective to solve the problem or for the implementation of a particular project.

The heterogeneity of the people involved in the group can be an enriching factor for the generation of brainstorming. During the process, no idea is discarded or taken as wrong or absurd. All ideas are accepted and treated for the definition of an effective solution. When you need quick answers to relatively simple questions, brainstorming is one of the most effective techniques.

There are three phases of brainstorming:

1. Identification of the facts:
    a. Definition of the problem
    b. Preparation of the elements
2. Generation of ideas
3. Finding the solution

It may be necessary to subdivide the problem into several parts—the brainstorming technique works for problems that have many possible solutions—then, it is necessary to collect information related to the problem. Next, ideas are generated through the brainstorming session and the solution is searched for by evaluating and selecting the best ideas.

Brainstorming is based on two principles and four basic rules.

The two principles are:

1. Delay in the trial of the ideas:
    - Most bad ideas are initially taken as good ideas. Delaying or deferring judgment provides the chance to generate a lot of ideas before settling on one. According to Osborn, human beings are capable of both judgment and creativity. Practicing a delay in judgment allows the creative mind to generate ideas without judging them. At first, it does not seem natural, but later, it has rewards. When ideas are generated, it is necessary to ignore considerations of the importance of the ideas, their usefulness, and their practicability. At this level, all ideas are equal. It is necessary to delay the trial until the generation of ideas is finished.
2. Creativity in quantity and quality:
    - The second principle relates to the quantity and quality of creativity. The more ideas are generated, the more likely finding a good idea is. Brainstorming takes advantage of the associations that are created when there are many ideas available. An idea can lead to another. Bad ideas can lead to good ideas.

Sometimes it is not possible to think about a problem until there are some answers. Brainstorming provides the possibility to put ideas that go through the minds of participants in the group on paper in order to achieve the best ones. Typically, guidelines, called rules, should be followed, although they are only guidelines.

The four basic rules of brainstorming are:

1. Rejecting criticism—This is probably the most important rule. Unless assessment is avoided, the principle of trial cannot operate. The group's failure to comply with this rule is the most critical reason a brainstorming session does not work. This rule is the one that primarily differentiates classic brainstorming from a meeting using traditional methods.
2. Encourage creativity—This rule is used to encourage participants to suggest any idea that comes to mind, without prejudice or fear that someone will assess it negatively. The most desirable ideas are those that initially seem more utopian and far from what appears to be a solution. It is important to have uninhibited participants as they generate ideas. When this rule is followed, it automatically creates an appropriate brainstorming climate, which also increases the number of ideas generated.
3. Encourage the amount of ideas—The more ideas are generated, the greater the chance of finding a good idea. In this case, the amount generates quality.
4. Establish a relationship between combination and improvement—The purpose of this rule is to encourage the generation of additional ideas for the construction and reconstruction of the ideas of others.

The following guidelines are usual in a brainstorming group:

- Explain the problem clearly
- Select a group of 2 to 10 participants
- Distribute text to participants about the problem
- Write the problem in a visible framework for all the group members
- Remind group members of the four basic rules outlined above
- Request new ideas of participants in the order in which they participate
- Request an idea from each participant, with only one idea in each round
- Record or write down all ideas
- Spend about 30 minutes per session, on average
- Select an evaluation group of three to five people
- Provide the evaluation group with the list of ideas, asking them to select the best ones
- Provide the original group a report with the ideas selected by the evaluation group
- Request the original group create a proposal for additional ideas stimulated by the previous list
- Give the final list of ideas to the working group

FIGURE 16.7
Brainstorming session.

Brainstorming groups are usually constituted of three types of elements (Figure 16.7):

1. The leader
2. The members
3. A secretary

People must be chosen who have some experience with the problem in question. The heads must not mix with the workers. People must be chosen who are at the same level of hierarchy in the organization. Most people cannot feel free or be creative enough when close to their boss.

The evaluation group must contain exactly three, five, or seven people (odd number). The reason for the use of an odd number is to eliminate the tie possibilities when the voting for possible solutions. The composition of the members of this group may vary. It may consist of people who were part of the group to generate ideas, a combination of people of this group with outsiders, or a completely new group of individuals.

The use of the same people has the advantage of ensuring familiarity with the problem, while the use of a group of people outside the original group has the benefit of greater objectivity.

16.10 Strengths, Weaknesses, Opportunities, and Threats Analysis

The term SWOT is an acronym for strengths, weaknesses, opportunities, and threats. The threats and opportunities are connected to the external organization's environment, while the weaknesses and strengths are connected to its internal environment.

There are no precise records on the origin of this type of analysis; however, it is attributed to two professors from Harvard Business School, Kenneth Andrews and Roland Christensen. However, its origin can be attributed to over 3000 years ago, through a quote from advice from Sun Tzu, 500 BC: "Focus on Strengths, recognize the Weaknesses, grasp Opportunities and take cover against Threats." Andrews (1971) introduced the SWOT concept in an in-depth way.

SWOT analysis is a very versatile tool for the analysis of many situations, such as a department of a company or the maintenance sector, regardless of size.

The primary function of SWOT analysis is to enable the choice of an appropriate strategy to achieve a certain goal through the critical evaluation of internal and external environments of the organization.

Figure 16.8 shows a scenario that divides the environment under analysis between internal (strengths and weaknesses) and external (opportunities and threats).

Strengths and weaknesses are determined by the current position of the company and almost always relate to internal factors. Opportunities and threats are anticipations of the future related to external factors.

The internal environment can be controlled by the top management, since it is the result of the action strategies defined by the governing board. Thus, during the analysis, whenever a strong point is identified, it must be emphasized. When a weakness is identified, the organization should seek to control it, or at least minimize its effects.

The external environment is totally out of the organization's control. But, despite this, it must be well known and monitored systematically in order to take advantage of opportunities and avoid threats. As this is not always possible, a plan can be made to address them, minimizing their effects. SWOT analysis should be used between the diagnosis and the formulation of the organization's strategy derived from it.

Regarding the filling of the SWOT frames, a possible approach is through the hierarchical structure shown in Figure 16.9.

SWOT analysis is usually qualitative. However, it may be supported on a quantitative basis in order to be able to characterize the strategy to be adopted more objectively. The method presented here shows a quantitative approach, following the next steps:

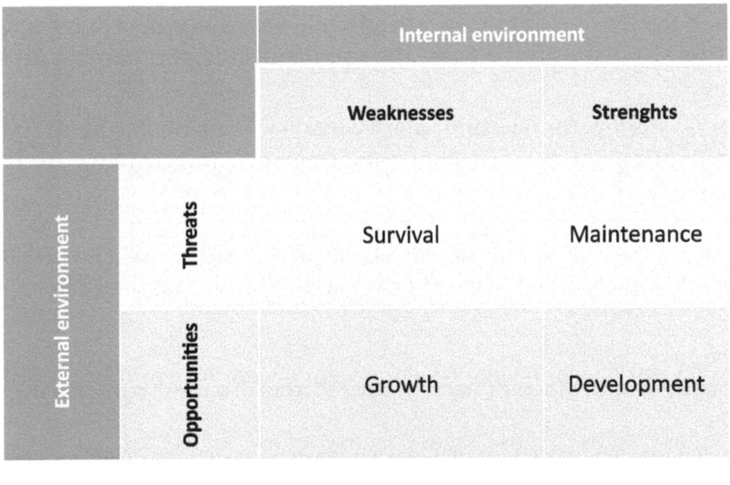

FIGURE 16.8
SWOT diagram.

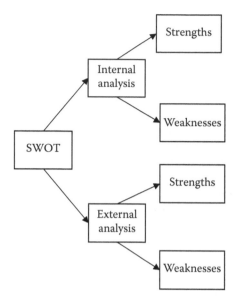

FIGURE 16.9
Hierarchical approach of SWOT matrix.

1. *Classification*
This consists of classification of the points observed in the internal and external environments in accordance with the requirements of the SWOT analysis. The rating considers relevant points of each one of the following:

   a. Threats—Refers to forces of the external environment, uncontrollable by the organization, that create obstacles to their strategy, but that may or may not be avoided if recognized in time

   b. Opportunities—Refers to forces of the external environment, uncontrollable by the company, that may support its strategic action, since they are recognized and used positively while they continue

   c. Strengths—Refers to internal structural advantages, controllable by the organization, that support it from the opportunities and threats of the external environment

   d. Weaknesses—Refers to structural disadvantages, controllable by the organization, that disadvantage it toward the opportunities and threats of the external environment

   Table 16.2 illustrates a table that allows structuring of the SWOT variables, leading to quantitative weighting together with other variables, as will be seen later.

2. *Quantification*
This consists of the importance of each analyzed requirement considering the following three categories:

   1. Very important—Weighting factor 4—Applies to cases where the requirement is fundamental in terms of impact on the organization's activities

TABLE 16.2

Structuring Table for SWOT Variables

Variable	Order Number	Item to Consider
Strengths	1	Strength 1
	2	Strength 2
	...	...
	n	Strength n
Weaknesses	1	Weakness 1
	2	Weakness 2
	...	...
	n	Weakness n
Opportunities	1	Opportunity 1
	2	Opportunity 2
	...	...
	n	Opportunity n
Threats	1	Threat 1
	2	Threat 2
	...	...
	n	Threat n

2. Important—Weighting factor 2—Applies to cases where the requirement has less relative importance in terms of impact on the organization's activities

3. Not very important—Weighting factor 1—Applies to cases where the requirements, regardless of the impact on the organization, are not reflected significantly in the subject under analysis, but should be considered in any case

The uniqueness of the weighting factors shall be 4, 2, and 1; this relates to their importance, with the choice of these coefficients made to safeguard the proportionality among them; that is, the previous category is 100% more important than the following category and reciprocally. Table 16.3 corresponds to Table 16.2 with the inclusion of the above classifications.

3. *Evaluation*
   - This consists of the intersection of the factors defined in the SWOT analysis in order to identify the most relevant aspects, as illustrated in Tables 16.4 through 16.6.
   - The evaluation of the factors makes up the sum of the products resulting from the previous evaluation in order to sequence the importance of the quadrants in descending order of score, which indicates the priority that the decisions should be given.

4. *Objectives for the creation of the strategy*
   - The last step is the analysis of the statement of goals for the most appropriate strategy to be adopted, which is defined as the sum of each quadrant (as indicated in the previous step), highlighting the quadrant that presents the highest score.

TABLE 16.3

Classification of SWOT Variables

Variable	Order Number	Item to Consider	Classification
Strengths	1	Strength 1	4, 2, or 1
	2	Strength 2	4, 2, or 1
	...	...	4, 2, or 1
	n	Strength n	4, 2, or 1
Weaknesses	1	Weakness 1	4, 2, or 1
	2	Weakness 2	4, 2, or 1
	...	...	4, 2, or 1
	n	Weakness n	4, 2, or 1
Opportunities	1	Opportunity 1	4, 2, or 1
	2	Opportunity 2	4, 2, or 1
	...	...	4, 2, or 1
	n	Opportunity n	4, 2, or 1
Threats	1	Threat 1	4, 2, or 1
	2	Threat 2	4, 2, or 1
	...	...	4, 2, or 1
	n	Threat n	4, 2, or 1

TABLE 16.4

Evaluation Matrix of Internal and External Factors

		External Analysis	
		Opportunities	Threats
Internal Analysis	Strengths	Demand high priority responses	Demand high priority responses
	Weaknesses	Opportunities that are not being taken advantage of	Weaknesses that must be transformed into strengths

The choice of strategy is made according to Table 16.7:

- Survival—For example, this refers to the adoption of measures to reduce costs, disinvestment (reduction of financial resources), or elimination of the activity
- Maintenance—For example, refers to the adoption of activity stability measures (specific focus on the range of activity) or expertise (focus on new technologies and new methodologies)
- Growth—For example, refers to the adoption of innovative solutions, the formation of partnerships, and related solutions
- Development—For example, refers to the approach of new courses of action, the provision of new services, the diversification of activity, and the development of their technological capacity

TABLE 16.5

Quantitative Evaluation Matrix

	Item	Weighting Factor	Opportunities				Threats			
			1	2	...	n	1	2	...	n
			4, 2, or 1	4, 2, or 1	4, 2, or 1	4, 2, or 1	4, 2, or 1	4, 2, or 1	4, 2, or 1	4, 2, or 1
Strengths	1	4, 2, or 1	*	*	*	*	*	*	*	*
	2	4, 2, or 1	*	*	*	*	*	*	*	*
	...	4, 2, or 1	*	*	*	*	*	*	*	*
	...	4, 2, or 1	*	*	*	*	*	*	*	*
	N	4, 2, or 1	*	*	*	*	*	*	*	*
Weaknesses	1	4, 2, or 1	*	*	*	*	*	*	*	*
	2	4, 2, or 1	*	*	*	*	*	*	*	*
	...	4, 2, or 1	*	*	*	*	*	*	*	*
	...	4, 2, or 1	*	*	*	*	*	*	*	*
	N	4, 2, or 1	*	*	*	*	*	*	*	*

*Is the result of the cross-product of the weighting factors of line × column.

TABLE 16.6

Quantitative Evaluation Matrix

		External Analysis	
		Opportunities	Threats
Internal Analysis	Strengths	Sum of product line × column of the quantitative evaluation matrix	Sum of product line × column of the quantitative evaluation matrix
	Weaknesses	Sum of product line × column of the quantitative evaluation matrix	Sum of product line × column of the quantitative evaluation matrix

TABLE 16.7

Strategy Matrix

		External Analysis	
		Opportunities	Threats
Internal analysis	Strengths	Development	Growth
	Weaknesses	Maintenance	Survival

16.11 Hoshin Kanri

The Hoshin Kanri method was developed in the 1960s by Japanese companies to manage the fulfillment of strategic objectives involving the whole functional structure of the organization. Its principle is that every department of the organization should incorporate a contribution to the overall objectives of the company in its management. Using this method brings a significant improvement in organizational performance by aligning the activities of all departments of the organization with its strategic goals.

The Hoshin Kanri method may be compared to the strategic asset management plan, which is the guide for setting asset management objectives. This last subject was described in this book, namely in Chapters 3 and 5.

16.12 A Case Study

The case study presented here focuses on the beverage industry in the sector of the stock of spare parts.

The process started with an audit of the maintenance state, as described in Chapter 4 of this book. This diagnosis identified the warehouse of the stock of spare parts as the weakest point of the maintenance management system.

As a consequence, the next step was the implementation of Lean management in the warehouse of the stock of spare parts, as will be described briefly next.

The parts warehouse, where maintenance materials are stored and managed, is a space with an area of approximately 250 m², located near the maintenance workshop and the maintenance management offices.

The main problem identified was the time necessary to find the spare parts in the warehouse. The sequence of operations is described in Figure 16.10.

Often, this task was complicated because of the high level of disorganization in the warehouse, including poor identification of the physical materials with their CMMS codes.

The solution adopted for the previous problem was the application of the 5S method and visual management in the spare parts warehouse.

Sectors in the plant were found where a very small number of parts were identified when compared to what was necessary, which implied that some interventions were pending or were done with minor quality.

For many others parts, there was a very high inventory value, which meant significant invested capital and resources consumed in their management. Obsolete materials were also identified, some of them due to technological obsolescence and others due to the fact that the equipment to which they belonged had already been removed from production lines.

The solution was to review the stock policy, analyze the stocks by Pareto analysis, apply the 5S methodology, and remove the obsolete materials with zero turnover.

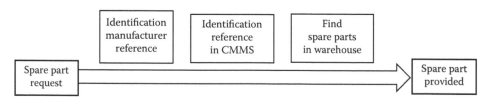

FIGURE 16.10
Sequence of operations to find spare parts in the warehouse.

It was found that the warehouse had several weaknesses, namely high disorganization, both by the sequence of the materials in the store and by the storage conditions, discriminated in the following way:

- No logical storage sequence
- Obsolete materials
- Materials that belong to line equipment that no longer exists
- Noninventoried materials
- Materials referenced with old labels in the CMMS
- Materials without assignment of location in the CMMS
- Materials dispersed throughout the warehouse
- Materials stored on the floor
- Materials exceeding the dimensions of boxes and shelves
- Several materials in the same box

Taking into account these situations, a plan was drawn up based on the following methodology:

- Layout and storage sequence:
  - Survey of existing layout
  - Elaboration of a new layout and storage sequence
- Visual management

  Visual identification is the most logical and fast way to find a part; for example, an electrical part, such as a contactor, where the easiest way to find it is through a visual sequence, like the following: Electrical Equipment => Contactors => Voltage Command => Contactor Power.

- Layout

  The warehouse has several shelves in which various storage boxes are placed. Each shelf corresponds to the materials of a production line. The sequence of lines currently installed in the factory starts at number 1 and ends at number 17. Initially, it was found that the sequence of the shelves was not in accordance with the sequence of the lines.

- Storage

  Some storage boxes had labels with outdated material designations and CMMS codes. In some cases, there were materials with CMMS codes, but not a label with their designation. The labels of the materials were changed to new labels, also including the physical location of the material.

- 5S + 1—Security
  - All materials that could have fallen and caused injury were transferred to horizontal locations where they were safely secured.
  - Although 5S is one basic Lean tool, it showed its high importance as an instrument that allowed the correction of several situations that had high impact on this process.

FIGURE 16.11
Initial view of the warehouse.

FIGURE 16.12
Final view of the warehouse.

Figures 16.11 and 16.12 show the initial view of the warehouse and the final view after the implementation of the Lean process.

17

Maintenance Standards

17.1 Background

Many standards have been very important to the maintenance field for the past decades. However, their use is not generalized around the world.

Because standards are not mandatory, their adoption is usually done by the most advanced organizations, usually for competitive reasons.

If the preceding happens in any economic field, in the ambit of physical assets in general, and in the maintenance area in particular, it assumes higher importance. The maintenance of facilities and equipment involves organization, management, and control. However, the organizational sector has been where the managers have taken a closer look.

However, today, some changes are happening: there are some transverse and specific maintenance standards that organizations can use in order to simultaneously fulfill the most exigent requirements for their assets and also to gain competitive advantages at the same time. On this subject, see also Farinha et al. (2013).

17.2 Portuguese Norm 4492 Maintenance Services Series

For many years, there was a lack in standards of service providers. Nowadays, with NP 4492:2010 (*Requirements for the provision of maintenance services*), that gap has been filled and standards can be used by maintenance service providers to obtain a certification, since they fulfill the requisites of the standard as previously referred to in Chapter 3; their main objectives are summarized in the following points:

- Define the requirements that maintenance service providers offer to their customers that are aligned with their needs and objectives. That is, they provide a guarantee of expected performance by keeping the asset operational and reliable, thereby reducing its downtime.
- Establish a benchmark for the certification of maintenance service providers and their periodic control by audits performed by an accredited entity.
- Support service providers by providing them with a resource for recognizing their efforts by distinguishing them from their competitors.
- Make the quality of maintenance services a permanent and transparent criterion for the customer, encouraging the implementation of the concept of life cycle cost

as a substitute for the acquisition cost, and a vector of commercial promotion and competitiveness for the maintenance service supplier.

- Encourage the establishment of a mechanism for self-regulation of the market itself, providing an increase in competence and innovation.

NP 4492:2010 has several standards that support it, corresponding to requisites that must be considered part of it, which are the following:

- NP 4483:2009, *Guide to the implementation of the maintenance management system*
- NP EN 13269:2007, *Maintenance—Instructions for preparation of maintenance contracts*
- NP EN 13306:2007, *Maintenance terminology*
- NP EN ISO 9000, *Quality management systems—Fundamentals and vocabulary* (ISO 9000: 2005)
- NP EN 13460:2009, *Maintenance—Maintenance documentation*
- NP EN 15341:2009, *Maintenance—Maintenance Performance Indicators*
- CEN/TR 15628:2007, *Maintenance—Qualification of Maintenance personnel*

Some standards are referred to using Portuguese nomenclature (NP—Portuguese Norm), but only NP 4492 and NP 4483 are Portuguese; the others are adapted from European Norms (ENs), from the International Organization for Standardization (ISO), and the European Committee For Standardization/Technical Report (CEN/TR). A summary of these standards is given next.

NP 4483:2009—*Guide to the implementation of the maintenance management system*

- This standard specifies the requirements for a maintenance management system that needs to demonstrate its ability to consistently provide a service that meets the customer requirements, including the applicable legal and regulatory requirements.
- The following norms contain information that constitutes requisites of this standard:
  - NP EN 13306:2007, *Maintenance terminology*
  - NP EN 13269:2007, *Maintenance—Instructions for preparation of maintenance contracts*
  - NP EN 13460:2009, *Maintenance—Maintenance documentation*
  - NP EN 15341:2009, *Maintenance—Maintenance Performance Indicators*
  - CEN/TR 15628:2007, *Maintenance—Qualification of Maintenance personnel*
  - NP 4492:2010, *Requirements for the provision of maintenance services*

EN 13269:2007—*Maintenance—Instructions for preparation of maintenance contracts*
The scope of this norm is to provide guidance for the preparation of maintenance service contracts. It can be applied to:

- Relationships between contractors and suppliers of maintenance services, both foreign and national
- The whole range of maintenance services, including planning, management, and control

- All kinds of equipment, with the exception of computer programs, unless the computer is subject to maintenance as an integral part of the technical equipment

To implement EN 13269, it is necessary to make use of the following standards:

- EN 13306:2007—*Maintenance terminology*
- EN 13460:2009—*Documents for maintenance*
- EN ISO 9000:2005—*Quality management systems—fundamentals for maintenance*

NP EN 13306:2007—*Maintenance terminology*

The scope of this norm is to specify generic terms and definitions for the technical, administrative, and management areas of maintenance. It is not applicable to terms used only for software maintenance.

NP EN ISO 9000—*Quality management systems—Fundamentals and vocabulary* (ISO 9000: 2005)

This standard describes the fundamental concepts and principles of quality management, which are universally applicable to:

- Organizations that aim to implement a quality management system
- Customers that seek confidence in an organization about its products and services
- Organizations seeking confidence in their supply-chain suppliers
- Organizations and interested parties seeking high-level relation quality management
- Organizations that aim to conform to assessments according to the requirements of ISO 9001
- Providers of training and related areas in quality management
- Developers of related standards

NP EN 13460:2009, *Maintenance—Maintenance documentation*

This standard specifies guidelines for:

- Technical documentation that must be provided with an asset before it is put into service to support its maintenance
- Information/documentation to be established during the operational phase of the asset to support its maintenance needs

This norm is primarily dedicated to designers, manufacturers, suppliers, technical writers of documentation, and documentation suppliers. It does not include documents related to training and skills of users, operators, and maintenance staff. It may not be applied to documentation of software maintenance. To implement this norm, it is necessary to make use of the following standards:

- EN 13269:2007—*Guideline on preparation of maintenance contracts*
- EN 13306:2007—*Maintenance terminology*
- EN 60300-3-14—*Dependability management—Part 3-14—application guide—maintenance and maintenance support* (International Electrotechnical Commission [IEC] 60300-3-14:2004)

NP EN 15341:2009—*Maintenance—Maintenance Performance Indicators*

The scope of this norm is to describe a management system of key performance indicators to measure the performance of maintenance under the influence of various factors, including economic, technical, and organizational ones. These indicators evaluate and aim to improve efficiency and effectiveness to help in the achievement of the maintenance excellence. To implement this norm, it is necessary to make use of the following standards:

- EN 13306:2007—*Maintenance terminology*
- IEC 60050-191 (International Electrotechnical Commission, 1990)—*IEC vocabulary, dependability and quality of service*

CEN/TR 15628:2007—*Maintenance—Qualification of Maintenance personnel*

The scope of this norm is to report the current situation for defining competence levels for personnel operating in the maintenance field and the knowledge levels required to be addressed to carry out those competencies.

The next norms are indispensable for the application of this standard:

- EN 13269:2007—*Guideline on preparation of maintenance contracts*
- EN 13306:2007—*Maintenance terminology*
- EN 13460:2009—*Documents for maintenance*
- NP EN 15341:2009, *Maintenance—Maintenance Performance Indicators*
- NP EN ISO 9000, *Quality management systems—Fundamentals and vocabulary* (ISO 9000: 2005)
- EN ISO 9001, *Quality management systems—Requirements* (ISO 9001:2000)
- EN ISO 14001, *Environmental management systems—Requirements with guidance for use* (ISO 14001:2004)
- IEC 60050-191 (International Electrotechnical Commission, 1990)—*IEC vocabulary, dependability and quality of service*

17.3 International Electrotechnical Commission 60300 Dependability Series

The International Electrotechnical Commission, through its technical committee TC56, develops and maintains the international standards that provide systematic methods and tools for dependability assessment and management of physical assets throughout their life cycles.

The standards prepared by IEC TC56 have been organized into three levels:

- Management
- Process
- Tools or supporting standards

IEC 60300, entitled *Dependability Management,* is the focal point of IEC TC56 and has the following parts:

- *IEC 60300-1: Dependability management. Part 1: Dependability programme management*
  - This standard, aimed at contracts, covers the requirements of dependability assurance and program elements. It supplies links between the supplier's and customers' organizations. It is possible to use this document separately or in conjunction with ISO 9001.
- *IEC 60300-2: Dependability management. Part 2: Dependability programme elements and tasks*
  - This standard contains requirements for establishing dependability programs covering reliability, maintainability, and maintenance support. It gives guidance to fit the design, development, and production processes to meet requirements and contractual conditions. It also provides a link between program elements and tasks.
- *IEC 60300-3: Dependability management. Part 3: Application guide*
  - This standard is designed to help users choose and apply the correct tools for a particular situation.

17.4 International Electrotechnical Commission 60812 Failure Mode and Effect Analysis

Failure mode and effect analysis and/or failure modes, effects, and criticality analysis are strategic tools to aid in physical asset maintenance. They are methodologies designed to identify potential failure modes for a product or process, assess the risk associated with those failure modes, rank the issues in terms of importance, and identify and carry out corrective actions to address the most serious concerns.

There are many published guidelines and standards for the requirements and recommended reporting format of FME(C)A. Some of these include SAE J1739, AIAG FMEA-3, and MIL-STD-1629A. In addition, many industries and companies have developed their own procedures to meet the specific requirements of their products/ processes.

IEC 60812 is another very important standard on FMEA analysis, having been prepared by IEC Technical Committee 56: Dependability—the same committee referred to in the previous section.

This second edition of this norm cancels and replaces the first edition published in 1985. The main changes from the previous edition are as follows:

- Introduction of the failure mode effects and criticality concepts
- Inclusion of methods used widely in the automotive industry
- Added references and relationships to other failure mode analysis methods
- Added examples
- Guidance on advantages and disadvantages of different FMEA methods

17.5 International Electrotechnical Commission 62278/European Norm 50126 Reliability, Availability, Maintainability, Safety

IEC 62278 and EN 50126 are identical and published by the European Committee for Electrotechnical Standardization (CENELEC) and IEC, respectively. The IEC 62278/EN 50126 standard deals with RAMS analysis, focused on railway transport, but it can be used in many other maintenance fields, which is the main reason to refer it in this section.

RAMS is defined in EN 50126 as the abbreviation of the terms reliability, availability, maintainability, and safety.

According to EN 50126, RAMS is a process or method that assists in the avoidance of failures already in the planning phase of projects.

These standards can also be used to reach the goal of having the assets work well, and because of this:

- These standards provide guidance to help increase confidence that the system guarantees the achievement of this goal.
- These standards describe how to define the targets in terms of reliability, availability, maintainability, and safety.
- These standards define a systematic process to demonstrate that these targets are achieved.
- These standards define the responsibilities within the RAMS process throughout the asset life cycle, that is, who is doing what in each phase of the asset life cycle.

RAMS analysis helps identify technical performance and safety at a system, module, or component level. Technical performance and safety are described by RAMS, which is increasingly important in many economic areas that are highly dependent on physical assets, namely in all sectors of industries with high investments and risks, because:

- RAMS analysis can be used during the development and implementation of new products or the planning and realization of new assets.
- RAMS management ensures the definition of systems, the performance of risk analysis, the identification of hazard rates, and detailed tests and safety certifications.
- RAMS may also include security, which means the protection of the system against external attacks.

EN 50126—*Railway applications—The Specification and Demonstration of Reliability, Availability, Maintainability and Safety (RAMS)* has several parts, namely:

- Part 1—Basic requirements and generic process
- Part 2—Systems approach to safety
- Part 3—Guide to the application of EN 50126-1 for rolling-stock RAMS
- Part 4—Functional safety: electrical/electronic/programmable electronic systems
- Part 5—Functional safety: software

As stated in the standard, RAMS characteristics for rolling stock (i.e., its long-term operating behavior performance), as for any other system (this extension makes its own standard, which represents its potential to be applied to any type of physical asset), form an important part of its overall performance characteristics.

In rolling-stock contracts, there is a great emphasis on the impact on end customers of service failures and on the economic and risk considerations of RAMS (the business perspective). Consequently, life cycle cost is used as a measure of customer satisfaction and to provide a wider perspective of RAMS importance in terms of business economics. The LCC approach represents a complete vision of the total cost of ownership. The contribution of RAMS to the LCC of rolling stock (and many physical assets) could be used to allow its economic evaluation.

17.6 Other Standards

There are many other standards that are used in the maintenance field, the so-called "vertical" norms. Some of them were referred to in Chapter 12, namely in Section 12.2.1 about vibration analysis and Section 12.2.2 about oil analysis.

Beyond the preceding areas, Chapter 12 refers to some others sensors/areas of condition monitoring, like the following:

- Chemical/gas
- Force/load/torque/strain
- Heat
- Humidity/moisture
- Motion/velocity/displacement/position
- Pressure
- Temperature sensors
- Water quality
- Voltage
- Current

For each of these variables, the specific norms and how to use them for each specific piece of equipment must be searched for. The main standard organizations around the world are the following:

- International Organization for Standardization—Within the ISO, there are specific technical committees dedicated to defining standards on specific issues, except those related to telecommunications engineering, which are the responsibility of the International Telecommunications Union (ITU) and electrical engineering, which is the responsibility of the International Electrotechnical Commission.
- International Telecommunications Union—Establishes telecommunications standards.
- International Electrotechnical Commission—Focuses on the standardization of electrical and electronic technologies. It is centered around the TC56 committee concerning dependability.

- Institute of Electrical and Electronics Engineers (IEEE)—Focuses on electrical and electronic technologies. One of the most important bodies is the computer committee Society Software Engineering Standards Committee (SESC).
- American National Standards Institute (ANSI) and National Institute of Standardization and Technology (NIST) from the United States—The Information Systems Conference Committee (ISCC) focuses on the development of IT standards, and the ANSI ISO Council (AIC) develops the relationship with ISO.
- European Conference of Posts and Telecommunications Administrations (CEPT)—An association of telecommunications companies, from which, in 1988, the European Telecommunications Standards Institute (ETSI) was created.
- Comité Européen de Normalisation (CEN)—Standardizes information and communication technology (ICT), concerned mainly with security of customers and the environment.
- Comité Européen de Normalisation Electrotechnique (CENELEC)—Centered on the definition of electrical engineering standards.
- British Standards Institution (BSI)—Publicly Available Specification (PAS) 99 is concerned with management systems. PAS 55 (optimal management of physical assets) is about life cycle management of capital investments, minimizing risks, and their integration according to ISO standards.
- Electronic Industries Association (EIA)—There are specific organizations in relation to:
  - G-33 for data management and configuration
  - G-34 concerning software
  - G-47 centered on systems engineering
- European Computer Manufacturers Association (ECMA)—It is an association of suppliers that, in cooperation with ISO, IEC, CEN and CENELEC, ETSI, and ITU, develops standards on ICT and consumption electronics.

18

Maintenance Project Management

18.1 Background

The program evaluation and review technique (PERT) and the critical path method (CPM) are complementary techniques that permit rationalization of project execution, that is, to rationalize the resources to reach the goals of a project, whether it is a new project, a renewal, or a big maintenance intervention for a physical asset with a long downtime before it becomes operational again.

PERT/CPM permits solution of the restrictions of the Gantt diagram, which consists of horizontal, parallel bars that indicate activities performed or to be performed, arranged in series on a horizontal time scale or arranged on top of each other, indicating concomitance of deadlines.

The biggest weakness of this technique is that it is impossible to represent the interdependence among different activities in the diagram, because the fact that some activities may be programmed for simultaneous periods does not necessarily make them interdependent.

18.2 Program Evaluation and Review Technique

PERT is a technique developed by D. G. Malcolm and others in a research and development program funded by the Office of Special Projects of the United States Navy, around 1958, to reduce the completion time of the Polaris Ballistic Projectile.

The purpose of PERT refers to tasks that are composed of activities never performed before; therefore, their duration is not precisely known. PERT is developed through a probabilistic model.

In a project, an activity is a task that must be performed. An event is a milestone marking the completion of one or more activities. Before an activity begins, all of its predecessor activities must be completed. The project network models represent the activities and milestones by arcs and nodes, respectively. The PERT chart may have many subtasks.

PERT planning involves the following steps:

1. Identify the activities and milestones.
2. Define the adequate sequence of activities.

3. Construct the network diagram.
4. Estimate the time required for each activity.
5. Determine the critical path.
6. Update the PERT network according to the project progress.

18.3 Critical Path Method

The CPM model is deterministic, which means that the data for its application is the project execution route, with the relation of dependence among events and the relation to the duration of each activity.

The CPM establishes the balance between the costs and date of finalization of the project. It addresses the rationalization of labor costs and other resources to find the appropriate duration of the project. The relationship between the variation of time for the accomplishment of the activities and the amount of resources allocated is known.

In the CPM, there are no uncertain times of achievement like in PERT. The CPM is particularly concerned with the time–cost relationship. The CPM is used in projects such as construction, equipment renovations, and annual plant shutdowns for maintenance activities.

18.4 Program Evaluation and Review Technique-Critical Path Method Networks

A PERT-CPM network permits:

- A graphic view of the activities that implement the project
- An estimate of the time the project will consume
- A vision of which activities are critical to reach the deadline to conclude the project
- A view of how much time is spent in noncritical activities and that can be discussed to reduce the application of resources and, consequently, their costs

The main variables to be considered to elaborate a PERT-CPM network are the following:

- Event—It is the milestone that marks the beginning or ending of an activity. In a project, the events are always presented as circles, which are numbered in ascending order according to the direction of the project progress.
- Activity—It represents the action that moves the tasks from one event to another, spending time and/or resources during the process. It is always represented by an arrow, oriented from the beginning to the end, without graphic scale.

The next figure (Figure 18.1) shows the relations between events and activities.

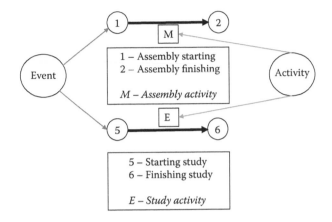

FIGURE 18.1
Relations between events and activities.

The steps to construct a PERT-CPM network are the following:

  i. Define and list the tasks to be carried out for project completion, that is, the activities themselves.
 ii. Define the preceding and subsequent tasks, that is, the activities' order of execution.
iii. Define the time execution of each task, that is, the duration of each activity.

The designer of a PERT/CPM network needs to list the activities that constitute the project and determine the interrelationships among them.
The rules to implement a PERT/CPM network are the following:

- Rule I
  - Each activity in the network is represented by one and only one arrow.
  - An activity may be broken down into smaller activities.
- Rule II
  - Two activities cannot be identified by the same final and initial event (Figure 18.2).
  - But, in practice, two activities can be performed simultaneously.

To overcome this problem, a fictitious activity is added to the network with a zero time consumption associated with it (Figure 18.3).

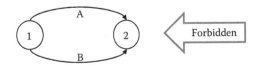

FIGURE 18.2
Two activities cannot be identified by the same final and initial event.

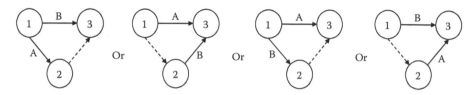

FIGURE 18.3
Fictitious activities.

- Rule III
  - To ensure the correct precedence relationship in the network, the following issues should be addressed when each activity is included in the network:
    - Which activities need to be finished immediately before the present activity can be started?
    - Which activities need to follow this activity?
    - Which activities need to occur simultaneously with this activity?

Only after the construction of the diagram are the events numbered. In this operation, the following points must be observed:

a. The number of the initial event of an activity must be less than that of the final event, including fictitious activities.
b. The number of the initial event is always 1 (one). The numbering must be continuous, following the sequence of the diagram, from left to right and from top to bottom, always following the preceding rule (a).

As a result of the numbering, according to this criterion, there is the alternative of the activities being referenced through the numbers of the initial and final events (which are unique for each activity): activity-> (initial event, final event).

There are some possible failures that may occur when a PERT-CPM network is designed, like the following:

a. No inclusion of activities
b. The relationship of interdependence is not well demonstrated
c. Nonexistence of interdependence
d. Unnecessary inclusion of fictitious activities
e. Errors in event enumeration

In a well-implemented network, it is mandatory that events and activities follow well-defined criteria.
An event should be:

- Specific and meaningful for the project
- Distinguishable at each moment
- Easily understandable to all project stakeholders

An activity should:

- Constitute a specific, tangible, and meaningful task
- Be designed in such a way that the responsibility for the work is identified
- Present an understandable description for all people
- Be executed within a well-defined period of time

To evaluate a PERT-CPM network, the following are necessary:

- The network evaluation permits definition of the project critical path.
- This evaluation is divided into two steps—advance and return:
  - Advance—In this step, the evaluations are made from the network initial node to the final node.
  - Return—In this step, the direction of the evaluations is the reverse.
- At each node or event, the following values are calculated:
  - Earliest event date
  - Latest event date

The preceding dates are the following:

- The earliest event date is the earliest date to start the activities that come from this event, counted from the beginning of the project, assuming that all activities that occur in this event were not delayed in their execution.
- The latest event date is the latest date to reach the event without delaying the project.

The values of the earlier and latest dates of the event are included in the network itself, next to the event number. A practical way to represent them is shown in Figure 18.4.

To calculate the earliest date of an event, the procedure is the following:

$$C(j) = \max_i[C(i) + D(i, j)] \tag{18.1}$$

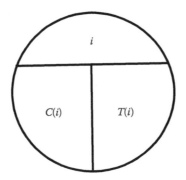

FIGURE 18.4
Representation of the earliest and latest dates.

where:

$C(i)$ = the earliest date of event i

$D(i, j)$ = the duration of the activity (i, j)

For the initial event of the project $i = 1$, the earliest date is always null, that is,

$$C(1) = 0$$

The synthesis of the procedure is the following:

- The counting is done from left to right, adding the durations of the tasks to each other and considering the highest value in the intersections—a task cannot start until all preceding tasks are completed.

To evaluate the calculation of the latest date of an event, the procedure is the following:

$$T(i) = \min_j [T(j) - D(i, j)] \tag{18.2}$$

where:

$T(j)$ = the latest date of event j

$D(i, j)$ = the duration of the activity (i, j)

For the latest project event $j = n$, the latest date is always equal to the earliest date of this event.

The synthesis of the procedure is the following:

- The counting is done from right to left, subtracting the durations of the tasks from one another, from the end date, and considering the lowest value at the intersections—a task cannot start later than one that permits completion of the project within the defined date.

After the PERT network is designed, it is time to evaluate the critical path and the time tolerances. To do this, it is necessary to start defining the first starting date and the last starting date.

- The first starting date (FSD)—This is the date to start the activity if the preceding activities started at the earliest opportunity and completed within the estimated duration, that is:

$$FSD(i, j) = C(i) \tag{18.3}$$

- The last starting date (LSD)—This is the start date of an activity so that the project does not suffer delays, that is:

$$LSD(i, j) = T(j) - D(i, j) \tag{18.4}$$

- The first completion date (FCD)—This is the end date considering that the activity starts at $FSD(i, j)$ and fulfils its estimated duration, that is:

$$FCD(i, j) = C(i) + D(i, j)$$

- The last ending date (LED)—This is the deadline for the completion of an activity, under penalty of delaying the project, that is:

$$LED(i, j) = T(j) \tag{18.5}$$

The total tolerance time (TTT) of an activity (i, j) can then be determined by the relations:

$$TTT(i, j) = LED(i, j) - FCD(i, j) \tag{18.6}$$

For the preparation of the project chronogram, it is necessary to know the activity dates and the time tolerance of its duration.

The chronogram is built on a frame where a horizontal scale indicates the evolution of time.

Initially, the critical activities are considered, including them as continuous lines in the schedule.

The noncritical activities are included in the timeline, indicating the first starting date and the last ending date for each activity as their implementation deadlines.

These boundaries are joined by dashed lines, indicating that these activities may have their execution programmed within this range, without any prejudice in the relations of precedence.

For each activity, there are also two continuous lines of length proportional to the duration of the activity:

- The first starts at the first starting date and, by construction, ends at the first completion date.
- The second continuous row starts at the last starting date and ends, therefore, at the last ending date.

The PERT-CPM network is a very important tool, with a lot of software tools, both open source and commercial. Some software for it includes the following:

- Critical Tools—project planning software—http://www.criticaltools.com/
- VERTEX42.com—http://www.vertex42.com/
- SmartDraw—http://www.smartdraw.com/
- Microsoft Project—http://www.microsoftstore.com/

18.5 Other Methodologies

PERT and the CPM have some limitations that make it very difficult to use them to model some complex projects. An alternative is the tool called the graphical evaluation and review technique (GERT), which includes features such as stochastic models, feedback loops, multiple outcomes, and repeat events.

The GERT features provide the capability to model and analyze projects and systems in a very general way.

For example, the GERT is a good tool for modeling and analysis when projects involve probabilistic occurrences, false starts, activity repetition, and multiple outcomes.

A GERT graph is built with one start node and some end nodes, which means different possibilities for project endings are possible.

The main drawback associated with the GERT is the use of Monte Carlo simulation required to model the GERT system. However, GERT networks have as strengths their graphical representation, which is intuitive and easy to understand.

In a GERT graph, the edges indicate tasks for which the project resources, such as time or costs, and their probability are allocated.

To create a GERT graph, it is necessary to follow the next steps:

i. Transform linguistic options of the project into a stochastic graph—Indicate project tasks and relations between them.

ii. Define the data regarding project tasks—Fix task durations or costs and their probability.

iii. Evaluate the substitute transformation—This gives information regarding the probability of project success and the expectations of total project time or costs.

iv. Interpret the results.

About this subject, please see Moore and Clayton (1976) and Kutschenreiter-Praszkiewicz (2017).

18.6 Case Study

The case presented here refers to a maintenance intervention to replace a component in equipment. Figure 18.5 presents the sequence of activities to implement it.

From the preceding sequence of activities, the first PERT network can be made, as is shown in Figure 18.6.

In the next step, the dependence relationships between activities are described (Figure 18.7).

Activity	Description	Dependency
A	Take the spare part (piece) from the warehouse	-
B	Turn off the equipment	A,C
C	Take the adequate tool	-
D	Withdraw the worn piece	B
E	Put the new piece into the equipment	D
F	Turn on the equipment	E
G	Put the worn piece in the container	D
H	Put the tool in its adequate place	E

FIGURE 18.5
Sequence of actions to change a piece in an equipment.

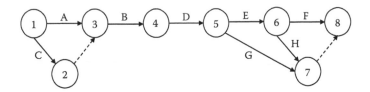

FIGURE 18.6
Initial PERT network for the maintenance actions.

Figure 18.8 shows the new PERT network, including the dependence relationships.

From Figures 18.8 and 18.7, which also include the duration of each activity, the PERT network can be constructed with the earliest and latest dates (Figure 18.9).

Then, the new table can be constructed with the times and respective margins, as can be seen in Figures 18.10 and 18.11, in which the critical path and the respective activities are emphasized.

Activity	Duration	Precedent Activities	Next Activities
A	3	-	D, E
B	6	-	G
C	2	-	H
D	4	A	G
E	2	A	H
F	7	A	-
G	4	B, D	-
H	3	C, E	-

FIGURE 18.7
Dependence relationship between activities.

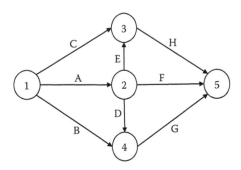

FIGURE 18.8
PERT network with the dependence relationships.

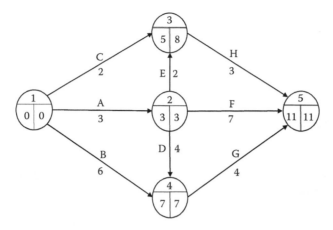

FIGURE 18.9
PERT network with the earliest and latest dates.

Activity	Duration	Starting Date		Finishing Date		Total margin interval
		FSD	LSD	FCD	LED	
A	3	0	0	3	3	0
B	6	0	1	6	7	1
C	2	0	6	2	8	6
D	4	3	3	7	7	0
E	2	3	6	5	8	3
F	7	3	4	10	11	1
G	4	7	7	11	11	0
H	3	5	8	8	11	3

FIGURE 18.10
Critical path—activities A-D-G.

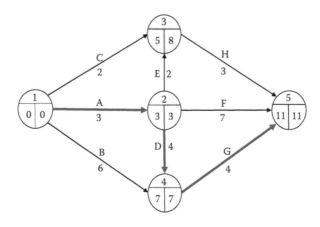

FIGURE 18.11
Critical path network—activities A-D-G.

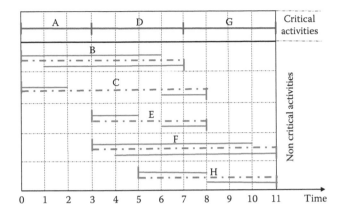

FIGURE 18.12
Project chronogram.

The inflexible activities are called critical, and the network that they form is called the critical path of the project. The critical path is the sequence of longer project activities. There is always at least one critical path in each project, but there may be several. A critical path is one in which activities have no margin to begin or to finish.

Finally, based on the last data, the chronogram is designed according to the rules described in the final part of the previous section (Figure 18.12).

19

Maintenance Training

19.1 Background

Maintenance activity requires that engineers and technicians be permanently up to date, which implies permanent training that can be fulfilled in several ways. Traditional training is done through knowledge communication between the teacher and the audience. This knowledge space can be a classroom, a laboratory, and/or training on site.

However, the new technological support, like professional videoconferences or similar common tools for free use on the Internet, such as devices for visual reality, mixed reality, and augmented reality (AR), permit a multiplicity of approaches that make the traditional ones seemingly obsolete.

But, similar to many other knowledge areas, in the future, solutions for maintenance training will be a mix of all the approaches referred to. Probably, a future learning space will blend both traditional classrooms and innovative online learning environments.

The next sections will present a synthesis of the most common methods used and those that may be used for maintenance training.

19.2 E/B-Learning

E-learning or e-learning corresponds to a teaching model based on technology, usually using dedicated channels (videoconference) or the Internet's capabilities for communication and content distribution (Figure 19.1). When these kinds of resources are used, the participants are usually in different locations using image and voice to communicate among themselves.

Training through e-learning can be synchronous or asynchronous:

- Synchronous teaching
  - The teacher and student are in class at the same time, and this can be done through chat, video conferencing, and web conferencing. This approach allows participants to ask questions and have discussions, making it possible for the teacher and students to be face to face at a distance.

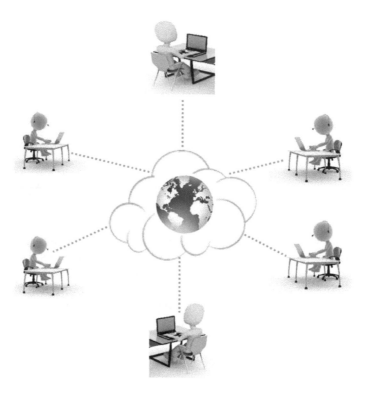

FIGURE 19.1
E-learning.

- Asynchronous teaching
  - The teacher and students are not in class at the same time, as is the case of email and forums. In corporate e-learning, many projects have no teacher because they were designed as self-training. The students enroll whenever they want, participate whenever they want, and end whenever they want. Asynchronous teaching distinguishes itself by its flexibility in the use of time—each student can take the course according to his or her own learning schedule.

Blended learning, or B-learning, refers to a training system where most of the content is transmitted online, usually through the Internet. The blended designation is also called mixed (Figure 19.2).

It can be structured as synchronous or asynchronous teaching in the same way as e-learning, that is, in situations where teacher and students work together at a predefined time or not, with each one performing tasks on defined schedules. However, blended learning in general is not totally asynchronous, because it would require an individualized availability to have face-to-face meetings.

FIGURE 19.2
B-learning.

19.3 Intelligent Learning Systems

An intelligent tutoring system is a computer system that aims to provide immediate and customized feedback to learners, usually without requiring intervention from a human teacher.

Intelligent tutoring systems have the goal of enabling training in a meaningful and effective way by using several computing technologies, like artificial intelligence and other up-to-date technologies.

There is a close relationship among intelligent tutoring, cognitive learning theories, and training design. An intelligent tutoring system usually aims to replicate the demonstrated benefits of one-on-one, personalized tutoring in contexts where the students would otherwise have access to one-to-many instruction from a single teacher (e.g., classroom lectures) or no teacher at all (e.g., e/B-learning) (Figure 19.3).

On this subject, please see Aquino et al. (2005), Palloff and Pratt (2007), Chen (2010), and Albarelli et al. (2013). These are only some references that also include information about the other sections of this chapter.

FIGURE 19.3
Intelligent learning systems.

19.4 Learning through Three-Dimensional Models

The approach to 3D models in asset management in general and in maintenance activity in particular was covered in Chapter 14, which demonstrates its importance
in this knowledge area and, as a consequence, the importance of learning through 3D
models.

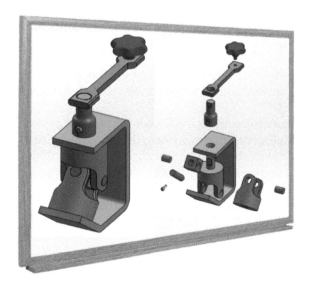

FIGURE 19.4
Learning through 3D models.

First of all, it is necessary to have 3D models of the facilities and equipment in order to make it possible to teach and learn through them. Usually, the main problem is that organizations do not have 3D models of their assets. Then, it is necessary to design them, which can be a very difficult task.

One way, probably the most usual, is to design the equipment piece by piece until the total asset is completely modeled. Another way, especially when the external view is the most relevant, is the use of a robot that scans a static scene using cameras and a 3D laser range scanner. In some situations, it may be interesting to make some physical 3D models using a 3D printer or Computer Numerical Control (CNC) equipment, including the most complex of five axes.

Figure 19.4 show a simple piece of equipment, completely assembled, and its pieces in assembly position, projected on a white board.

19.5 Learning through the Use of Sensors

The use of sensors in games created by the principal global manufacturers, permitting players to interact virtually with scenarios, opens up new possibilities for training, especially in the maintenance field.

In fact, it is possible to design tools based on sensors that detect the position of the human body, including the correct position of the head, legs, and arms. Thus, it is possible to make the learner interact with the equipment, assembling and disassembling it and performing specific operations, including the solution of faults (Figure 19.5).

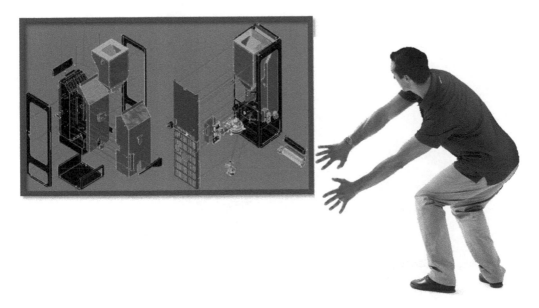

FIGURE 19.5
Learning through the use of sensors.

19.6 Learning through Virtual Reality

Virtual reality and mixed reality are new real technologies that must be considered today and in the near future as new ways to help the process of training of maintenance professionals and physical asset managers.

This new reality can be, as of this writing, implemented as a virtual and or mixed reality campus that can be composed by 3D modeling tools, and virtual reality with head-mounted displays (HMDs).

Based on this virtual campus, users will be able to explore this virtual space using several interfaces, like a traditional desktop or a full 3D immersive device.

On this campus, webcams, panoramic 360° video, and holograms can create virtual training scenarios in the real-world classroom (Figure 19.6).

Virtual reality has the potential to combine the best aspects of both real-world classrooms and online distance learning into a single platform. Virtual reality tools can permit the simulation of virtual people who represent the teachers and students, including the use of voice and video capabilities, slide projection, and other collaborative technologies. This whole new learning world can be implemented in a friendly and cost-effective environment, including for distance learning.

Training becomes really interactive using virtual scenery. Virtual-world environments also add the enormous advantage of permitting engagement with other learners, even at distance.

FIGURE 19.6
Learning through virtual reality.

Virtual learning opens up the possibility of conjugating a lot of technological tools added to the conventional ones to give access to the state of the art about maintenance knowledge in particular and asset management in general.

19.7 Learning through Augmented Reality

AR is a technology that enriches the real environment by superimposing virtual data on it. There are a lot of variations of AR systems, starting with hardware: the virtual contents displayed may be 2D data or 3D models, human-machine interaction, or even the way in which the target locations are identified in a scene (Figure 19.7). About this subject, please see a project supervised by the author in Oliveira et al. (2014).

Augmented reality is emphasized here because of its relevance in the maintenance field. It has the potential to help significantly decrease the time necessary to perform a maintenance intervention, whether planned or necessary to correct a fault. Additionally, it has the potential to permit any technician to solve a fault in equipment through an adequate AR application, even if the technician is not the one who usually deals with the equipment with the fault.

The results from several projects in different activity sectors, like aerospace, automotive, defense, and health, show that AR achieves promising results and is a powerful technology to support technicians.

However, AR still has a lot of restrictions, especially the use of markers. In fact, object detection and tracking in nonlaboratory environments is a challenging task. This is why, usually, the identification of parts is done based on AR markers. On the other hand, markers represent a restrictive and impractical solution for many situations. For this reason, object markerless identification is one of the most active subjects in terms of applied research on AR.

To create an AR-based solution, it is necessary to have all components modeled in 3D and the procedures prepared so they can be shown to the trainee. For maintenance training,

FIGURE 19.7
Learning through augmented reality.

it is easier to create AR solutions when CAD models are available from the manufacturer. For assets that do not have these models, they must be designed before implementing the AR solution.

An additional capacity that can be added to AR systems used in maintenance activity, especially to aid fault diagnosis, is artificial intelligence. These two tools in conjunction, AR and AI, can aid the technician in solving faults efficiently and quickly. If the technician uses AR glasses, those tools can be added to an interface controlled by voice, aiming to permit the technician to have both hands free to perform the intervention.

20

Terology beyond Tomorrow

A book is always the beginning of a trip and never ends. This means creating expectations about what will come next.

The asset management and engineering methodologies that support this book are in a state of accelerated evolution. Because of this, it is not easy to know when it will stop or when this movement will reduce a level of creation and incorporation of new technologies and methodologies to aid in the organization and management of physical assets.

Regarding management itself, the introduction and certification of many organizations by ISO 55000 will be a revolution when compared with the current vision of asset management. The word "cost" will be replaced by the asset's fixed initial investment and the variable asset investment over time.

The relevance of assets in companies added to the influence of technologies associated with Industry 4.0, which makes assets more and more intelligent, will be the next future reality.

The next future technology to aid asset management will use the current consolidated support in an integrated and transparent way: artificial intelligence, big data, the Internet of Things, the Internet of People, cloud computing, visual reality, and augmented reality, among others.

However, some of these technologies will evolve, especially through greater standardization that today almost does not exist. When this is a reality, EAM and CMMS systems will be more portable and will communicate transparently. This means that intelligent sensors will communicate through standard protocols that will permit connection of any sensor to any system independently of its manufacturer or operating system (OS).

In the future, augmented reality systems will work without markers and will be basic tools for any technician. This technology will work using voice commands, with natural language, and associated with artificial intelligence, and will permit manipulation of 3D models of each piece of equipment to aid interventions.

In addition to the new AR systems, holographic systems that will work with 3D models will appear. This will permit manipulation in space of the 3D equipment's components, simulating the best maintenance solutions, especially in the case of faults, aided by AI tools, as previously mentioned.

Terology behind tomorrow will be a new reality in all extensions of this concept, where physical assets will have a new, more pragmatic vision and a new role in companies. This will permit increased profits through more pragmatic management of physical assets and a more pragmatic and respected role for the industrial engineer in particular and maintenance professionals in general.

References

Adams, M. L. 2009. *Rotating Machinery Vibration: From Analysis to Troubleshooting*. USA: CRC Press; 2nd Edition. ISBN-10: 1439807175. ISBN-13: 978-1439807170.

Aksel A. T., Henrik, S.-O., Gorm, J., Jens, T., Erik, K. 2010. A Robotic Concept for Remote Maintenance Operations: A Robust 3D Object Detection and Pose Estimation Method and a Novel Robot Tool. *The 2010 IEEE/RSJ International Conference on Intelligent Robots -and Systems*. October 18–22, 2010, Taipei, Taiwan. Pp. 5099–5106.

Albarelli, A., Bergamasco, F., Celentano, A., Cosmo, L., Torsello, A. 2013. Using multiple sensors for reliable markerless identification through supervised learning. *Machine Vision and Applications* 24(7):1539–1554. doi: 10.1007/s00138-013-0492-2.

Andrews, J., Dugan, J. 1998. *Fault Tree Analysis: For Reliability and Risk Assessment*. USA: John Wiley & Sons Ltd. ISBN-10: 0471982792. ISBN-13: 978-0471982791.

Andrews, K. R. 1971. *The Concept of Corporate Strategy*. USA: Dow Jones-Irwin. ISBN-10 0870940120, ISBN-13 978-0870940125.

Angeli, C. 2010. Diagnostic expert systems: From expert's knowledge to real-time systems. *TMRF e-Book. Advanced Knowledge Based Systems: Model, Applications & Research*, Sajja & Akerkar (Eds.), Vol. 1, Pp. 50–73.

Assis, R. 2010. *Apoio à Decisão em Manutenção na Gestão de Activos Físicos*. s.l.: Edições Lidel, Portugal, 2010. 9789727576050.

Banks, J. 2009. *Discrete-Event System Simulation*. USA: Prentice Hall. ISBN 0136062121, 9780136062127.

Basharin, G. P., Langville, A. N., Naumov, V. A. 2004. The life and work of A.A. Markov. *Linear Algebra and Its Applications* 386:3–26. www.elsevier.com/locate/laa. Available online at www.sciencedirect.com.

Bazaraa, M. S., Sherali, H. D., Shetty, C. M. 2006. *Nonlinear Programming: Theory and Algorithms*. New Jersey: Wiley-Interscience; 3rd Edition. ISBN-10: 0471486000. ISBN-13: 978-0471486008.

Bellamine, M., Abe, N., Tanaka, K., Taki, H. 2002. Remote Machinery Maintenance System with the Use of Virtual Reality. *Proceedings of the First International Symposium on 3D Data Processing Visualization and Transmission (3DPVT.02)*, Padova, Italy. IEEE Computer Society. ISBN 0-7695-1521-5/02. doi: 10.1109/TDPVT.2002.1024036.

Bellman, R. 2003. *Dynamic Programming*. Dover Books on Computer Science. New York: Dover Publications. ISBN-10: 0486428095. ISBN-13: 978-0486428093.

Ben-Daya, M., Duffuaa, S. O., Raouf, A., Knezevic, J., Ait-Kadi, D. 2009. *Handbook of Maintenance Management and Engineering*. London, New York: Springer. ISBN-10: 1848824718, ISBN-13: 978-1848824713.

Bershad, A. 1991. Staffing and productivity assessment of the engineering department. *Health Estate Journal* 45(5):7–10, 12–14, 16–21.

Bertsekas, D. 2016. *Nonlinear Programming*. Series: Optimization and Computation. Massachusetts: Athena Scientific; 3rd Edition. ISBN-10: 1886529051. ISBN-13: 978-1886529052.

Bethune, J. D. 2016. *Engineering Design Graphics with Autodesk Inventor 2017*. USA: Peachpit Press. ISBN-10: 0134506979. ISBN-13: 978-0134506975.

Billinton, R., Allan, R. N. 1992. *Reliability Evaluation of Engineering Systems – Concepts and Techniques*. New York: Springer. ISBN 978-1-4899-0687-8. doi: 10.1007/978-1-4899-0685-4.

Birge, J. R., Louveaux, F. 2011. *Introduction to Stochastic Programming*. Springer Series in Operations Research and Financial Engineering. New York: Springer; 2nd Edition. ISBN-10: 1461402360. ISBN-13: 978-1461402367.

Bogue, R. 2013. Sensors for condition monitoring: A review of technologies and applications. *Emerald. Sensor Review* 33(4), 295–299. ISSN: 0260-2288.

Bonabeau, E., Theraulaz, G., Dorigo, M. 1999. *Swarm Intelligence: From Natural to Artificial Systems*. Santa Fe Institute Studies on the Sciences of Complexity. New York: Oxford University Press. ISBN-10: 0195131592. ISBN-13: 978-0195131598.

Bondy, A., Murty, U. S. R. 2010. *Graph Theory (Graduate Texts in Mathematics)*. New York: Springer. ISBN-10: 3642142788. ISBN-13: 978-3642142789.

Cappé, O., Moulines, E., Ryden, T. 2010. *Inference in Hidden Markov Models*. New York: Springer. ISBN-10: 1441923195. ISBN-13: 978-1441923196.

Chen, C. J. 2010. *Virtual Reality (VR)-Based Learning Environment: Design, Develop, Evaluate*. Germany: LAP LAMBERT Academic Publishing. ISBN-10: 3838354192. ISBN-13: 978-3838354194.

Ching, W.-K. 2009. *Markov Chains: Models, Algorithms and Applications*. Springer US. ISBN-10: 1441939865. ISBN-13: 978-1441939869.

Ciang, C. C., Lee, J.-R., Bang, H.-J. 2008. Structural health monitoring for a wind turbine system: A review of damage detection methods. *Measurement Science and Technology* 19(12):1–20. doi: 10.1088/0957-0233/19/12/122001.

Cobzaru, M. 2002. *Tutorial on Performance Metrics for Intelligent Systems*. Prepared for: Agent Based Software Engineering SENG 609.22, Dr. Behrouz H. Far. University of Calgary.

Conforti, M., Cornuéjols, G., Zambelli, G. 2014. *Integer Programming (Graduate Texts in Mathematics)*. New York: Springer; 2014 Edition. ISBN-10: 3319110071. ISBN-13: 978-3319110073.

Costa, J. A. M. 2002. Uma Abordagem ao Diagnóstico do Estado da Manutenção em Empresas Industriais. FEUP, Porto, Portugal, *MSc thesis*.

Costa, J. A. M., Farinha, J. T., Vasconcelos, B. C. 2000. Diagnóstico da Manutenção – Um Passo Fundamental na Sua Reorganização (Maintenance Diagnostics – A Fundamental Step in Your Reorganization). *Revista MANUTENÇÃO (Maintenance Magazine)* 65, 2° Trim, 14–18.

de Aquino, M. S., de Sousa, F. F., Frery A. C. 2005. Three-Dimensional Virtual Environments Adaptive to the Student's Profile for Distance Learning. *XVI Simpósio Brasileiro de Informática na Educação—SBIE—UFJF*. Pp424–433.

Deo, N. 2016. *Graph Theory with Applications to Engineering and Computer Science*. USA: Dover Publications; Reprint Edition. ISBN-10: 0486807932. ISBN-13: 978-0486807935.

Diaz, M. 2009. *Petri Nets: Fundamental Models, Verification and Applications*. UK and USA: Wiley-ISTE. ISBN-10: 1848210795. ISBN-13: 978-1848210790.

DoD 2011. *Department of Defense – Manual – Subject: Reliability Centered Maintenance (RCM)*. NUMBER 4151.22-M. June 30.

Dorigo, M., Stützle, T. 2004. *Ant Colony Optimization*. USA: MIT Press. A Bradford Book. ISBN-10: 0262042193. ISBN-13: 978-0262042192.

Farinha, J. M. T. 1994. *Uma Abordagem Terológica da Manutenção dos Equipamentos Hospitalares*. FEUP, Porto, Portugal, *Tese de Doutoramento*.

Farinha, J. M. T. 1997. *Manutenção das Instalações e Equipamentos Hospitalares*. Coimbra: Livraria Minerva Editora. ISBN: 9728318162.

Farinha, J. M. T. 2011. *Manutenção – A Terologia e as Novas Ferramentas de Gestão*. Lisboa; 1ª Edição, Monitor – Projecto e Edições, Lda.

Farinha, J. T. 2009. The Contribution of Terology for a Sustainable Future. *Proceedings of the 3rd WSEAS Int. Conf. on ENERGY PLANNING, ENERGY SAVING, ENVIRONMENTAL EDUCATION*. University of La Laguna, Tenerife, Canary Islands Spain, July 1–3. ISSN: 1790-5095; ISBN: 978-960-474-093-2. Pp. 110–118.

Farinha, J. T., Fonseca, I., Marques, V., Brito, A., Marimba, A., Pincho, N., Simões, A. 2004. A global view of maintenance management: From maintenance diagnosis to know-how retention and sharing. *WSEAS Transactions on Systems* 3(4):1703–1711, June; ISSN 1109-2777. Pp. 1703–1711.

Farinha, J. T., Fonseca, I., Simões, A., Barbosa, F. M., Viegas, J. 2008. New ways for terology through predictive maintenance in an environmental perspective. *WSEAS Transactions on Circuits and Systems* 7(7):July, 630–647. ISSN 1109-2734.

Farinha, J. T., Fonseca, I., Simões, A., Costa, A., Bastos, P., Barbosa, F. M., Ferreira, L. A., Carvas, A. 2010. Terology beyond tomorrow. *Maintworld – Maintenance & Asset Management* 1:46–50. ISSN 1798-7024, ISSN-L 1798-7024.

Farinha, J. T., Galar, D., Fonseca, I., Kumar, U. 2013. Certification of maintenance providers: A competitive advantage. *Journal of Quality in Maintenance Engineering* 19(2):144–156. doi: 10.1108/13552511311315959.

Ferreira, L. A. 1998. *Uma Introdução à Manutenção*. Porto, Portugal: Publindústria, Edições Técnicas. ISBN 972-95794-4-X.

Fonseca, I., Farinha, J. M., Barbosa, F. M. 2014. Maintenance planning in wind farms with allocation of teams using genetic algorithms. *IEEE Latin America Transactions* 12(6):1062–1070. ISSN: 1548-0992. doi: 10.1109/TLA.2014.6894001.

Giesecke, F. E., Mitchell, A., Spencer, H. C., Hill, I. L., Dygdon, J. T., Novak, J. E., Loving, R. O.; Lockhart, S., Johnson, C. 2016. *Technical Drawing with Engineering Graphics*. USA: Peachpit Press; 15th Edition. ISBN-10: 0134306414. ISBN-13: 978-0134306414.

Girault, C., Valk, R. 2002. *Petri Nets for Systems Engineering*. New York: Springer. ISBN-10: 3540412174. ISBN-13: 978-3540412175.

Goldman, S. 1999. *Vibration Spectrum Analysis*. New York: Industrial Press, Inc. ISBN-10: 0831102152. ISBN-13: 978-0831102159.

Gopalakrishnan, P., Banerji, A. K. 2004. *Maintenance and Spare Parts Management*. India: Prentice-Hall of India Pvt.Ltd. ISBN-10: 8120306694. ISBN-13: 978-8120306691.

Hastings, N. A. J. 2015. *Physical Asset Management: With an Introduction to ISO55000*. Springer; 2nd Edition. ISBN 978-3-319-14776-5 ISBN 978-3-319-14777-2 (eBook). doi: 10.1007/978-3-319-14777-2.

Haviv, M. 2013. *Queues: A Course in Queueing Theory*. Springer. ISSN 0884-8289. ISBN 978-1-4614-6764-9. ISBN 978-1-4614-6765-6 (eBook). doi: 10.1007/978-1-4614-6765-6.

Heisler, R. 2003. *Planning and Scheduling in a Lean Maintenance Environment*. Life Cycle Engineering, Inc. www.lce.com.

Herrera, C. N. B. 2017. *Stochastic Programming: Theory, Applications and Impacts*. Nova Science Pub Inc. ISBN-10: 1536109401. ISBN-13: 978-1536109405.

Heuser, C. A. 1991. *Conceptual Modeling of Systems* EBAI—IV Brazilian-Argentine Computer Science School, Brazil.

Houten, F. J. A. M. van., Kimura, F. 2000. *The virtual maintenance system: A computer-based support tool for Robust design, product monitoring, fault diagnosis and maintenance planning*. CIRP Annals: Manufacturing Technology 49(1):91–94. doi: 10.1016/S0007-8506(07)62903-5.

Husband, T. M. 1976. *Maintenance Management and Terotechnology*. Westmead, England: Saxon House. ISBN: 0566001462.

Ikegami, T., Shimura, T., Koike, M. 2001. *Plant Life Management and Maintenance Technologies for Nuclear Power Plants*. Hitachi. http://www.hitachi.com/ICSFiles/afieldfile/2004/06/08/r2001_03_105.pdf

Juran, J. M. 1992. *Juran on Quality by Design: The New Steps for Planning Quality into Goods and Services*. USA: Free Press; Revised Edition, ISBN-10: 0029166837, ISBN-13: 978-0029166833.

Kelly, S. G. 2006. *Advanced Vibration Analysis (Mechanical Engineering)*. CRC Press. ISBN-10: 0849334195. ISBN-13: 978-0849334191.

Kioskea.net. n.d. Kioskea.net; http://pt.kioskea.net/contents/projet/cahier-des-charges.php3; Accessed June 6, 2015, de Kioskea.net:

Kume, H. 1985. *Statistical Methods for Quality Improvement*, AOTS, Tokyo. ISBN – 4-906224-34-2.

Kutschenreiter-Praszkiewicz, I. 2017. Graph theory in product development planning. *Mechanisms and Machine Science*. S. Zawiślak & J. Rysiński (Eds.), Vol. 42, Pp. 165–173. Switzerland: Springer, ISBN 978-3-319-39018-5. doi: 10.1007/978-3-319-39020-8.

Levitt, J. 2008. *Lean Maintenance*. Industrial Press, Inc. ISBN-10: 083113352X. ISBN-13: 978-0831133528.

Lin, Z., Pearson, S. 2013. *An Inside Look at Industrial Ethernet Communication Protocols*. Texas Instruments. November 2013.

Makridakis, S. G., Wheelwright, S. C. 1989. *Forecasting Methods for Management*. John Wiley & Sons Inc. ISBN-10: 0471600636, ISBN-13: 978-0471600633.

Makridakis, S. G., Wheelwright, S. C., Hyndman, R. J. 1997. *Forecasting: Methods and Applications*. Wiley; 3rd Edition. ISBN-10: 0471532339; ISBN-13: 978-0471532330.

Marranghello, N. 2005. *Redes de Petri: Conceitos e Aplicações*. DCCE/IBILCE/UNESP; UNESP - Universidade Estadual Paulista "Júlio de Mesquita Filho".

Marques, V. 2005. *Diagnóstico de avarias em equipamentos baseado em informação difusa oriunda dos técnicos de manutenção.* Faculdade de Engenharia da Universidade do Porto, Tese de Doutoramento.

Marques, V., Farinha, J. T., Brito, A. 2001. *Diagnóstico de Falhas e Sistemas Periciais—Tópicos sobre o estado da arte.* Revista Manutenção, 68, 1º Trimestre de.

Marques, V., Farinha, J. T., Brito, A. 2009. Case-based reasoning and fuzzy logic in fault diagnosis. *WSEAS Transactions on Computers* 8(8):1408–1417. ISSN: 1109-2750.

May, V. V., Christensen, A. 2015. *3-D Engineering: Design and Build Your Own Prototypes (Build It Yourself).* Nomad Press. ISBN-10: 1619303116. ISBN-13: 978-1619303119.

Moczulski, W., Przystałka, P., Wachla, D., Panfil, W. 2013. *Interactive Education of Engineers in the Field of Fault Diagnosis and Fault-Tolerant Control.* Pomiary Automatyka Robotyka nr 11/2013. Pp84–91.

Mohanty, A. R. 2014. *Machinery Condition Monitoring: Principles and Practices.* CRC Press. ISBN-10: 1466593040. ISBN-13: 978-1466593046.

Moore, L. J., Clayton, E. R. 1976. *GERT Modeling and Simulation: Fundamentals and Applications.* Krieger Publishing Co., Inc., Melbourne, FL, USA. ISBN:0884053288.

Moubray, J. 1997. *Reliability-Centered Maintenance.* Industrial Press, Inc. ISBN-10: 0831131462. ISBN-13: 978-0831131463.

Norris, J. R. 1998. *Markov Chains.* Cambridge University Press. Cambridge Series in Statistical and Probabilistic Mathematics. ISBN-10: 0521633966, ISBN-13: 978-0521633963.

Nunes, M. 2012. *MANVIA – SGS certifica gestão da manutenção pela NP 4492.* Faxinforme. 1-05-2012.

Oliveira, R., Farinha, J. T., Raposo, H., Pires, N. 2014. Augmented Reality and the Future of Maintenance. *Proceedings of MPMM2014.* Coimbra, Portugal. ISBN 978-972-8954-42-0. http://dx.doi.org/10.14195/978-972-8954-42-0_12. Pp. 81–88.

Olmos, J. 1979. *A Modular Approach to Fault Tree Analysis.* University of Michigan Library. ASIN: B0030ZSDCY.

Orsburn, D. K. 1991. *Spares Management Handbook.* McGraw-Hill. ISBN-10: 0830676260. ISBN-13: 978-0830676262.

Osada, T. 1991. *The 5S's: Five Keys to a Total Quality Environment.* Quality Resources. ISBN-10: 9283311167. ISBN-13: 978-9283311164.

Osborn, A. F. 1963. *Applied Imagination: Principles and Procedures of Creative Problem-Solving.* Charles Scribner's Sons. ASIN: B000H5HJBQ.

Palloff, R. M., Pratt, K. 2007. *Building Online Learning Communities: Effective Strategies for the Virtual Classroom.* Jossey-Bass. ISBN-10: 0787988251. ISBN-13: 978-0787988258.

Paris, Q. 2016. *An Economic Interpretation of Linear Programming.* Palgrave Macmillan. ISBN: 978-1-137-57391-9. E-PDF ISBN: 978-1-137-57392-6. doi: 10.1057/9781137573926.

Petridis, V., Kehagias, A. 1998. *Predictive Modular Neural Networks—Applications to Time Series.* Springer Science+Business Media. ISBN 978-1-4613-7540-1.

Pincho, N., Marques, V., Brito, A., Farinha, J. T. 2006. E-learning by experience: How CBR can help. *WSEAS Transactions on Advances in Engineering Education* 3(7):699–704. ISSN 1790-1979. ISBN~ISSN: 1790-5117, 960-8457-53-X.

Plant, R. T., Salinas, J P. 1994. Expert systems shell benchmarks: The missing comparison factor. *Information & Management* 27: 89–101.

Popova-Zeugmann, L. 2013. *Time and Petri Nets.* Springer. ISBN 978-3-642-41114-4.

Pouliot, N., Montambault, S. 2008. Geometric Design of the LineScout, a Teleoperated Robot for Power Line Inspection and Maintenance. *2008 IEEE International Conference on Robotics and Automation.* Pasadena, CA, USA, May 19–23, 2008. Pp. 3970–3977.

QS 9000, Manuais da 1997. *Análise de Modo e Efeitos de Falha Potencial (FMEA). Manual de Referência.* Chrysler Corporation, Ford Motor Company, General Motors Corporation.

Rabiner, L. R. 1989. A Tutorial on Hidden Markov Models and Selected Applications in Speech Recognition. *Proceedings of the IEEE* 77(2):257–286.

Raposo, H., Farinha, J. T., Fonseca, I., Ferreira, L. A., Armas, F. 2012. A New Approach for the Diagnosis of State Maintenance. *2nd International Workshop and Congress on eMaintenance, Organized by Division of Operation and Maintenance Engineering, Process IT Innovations, from*

Luleå University of Technology, Sweden. Proceedings of eMaintenance 2012, Luleå, 12–14 December. Pp. 229–234.

Reisig, W. 2013. *Understanding Petri Nets Modeling Techniques, Analysis Methods, Case Studies.* Berlin: Springer. ISBN 978-3-642-33277-7.

Robert, P. 2013. *Stochastic Networks and Queues.* Series, Stochastic Modelling and Applied Probability (Book 52). Springer. ISBN-10: 3642056253. ISBN-13: 978-3642056253.

Saeed, R. A., Galybin, A. N., Popov, V. 2013. 3D fluid – structure modelling and vibration analysis for fault diagnosis of Francis turbine using multiple ANN and multiple ANFIS. *Elsevier. Mechanical Systems and Signal Processing* 34:259–276.

Scheffer, C., Girdhar, P. 2004. *Practical Machinery Vibration Analysis and Predictive Maintenance.* Newnes; 1st Edition. Paperback ISBN: 9780750662758. eBook ISBN: 9780080480220.

Schrijver, A. 1998. *Theory of Linear and Integer Programming.* Wiley. ISBN-10: 0471982326. ISBN-13: 978-0471982326.

Seet-Larsson, K. G. 2010. *A Cookie Cutter Introduction to FMEA and FMECA: A Practical Example from Theory to Implementation.* Lap Lambert Academic Publishing. ISBN-10: 383836905X. ISBN-13: 978-3838369051.

Shih, R. 2016. *AutoCAD 2017 Tutorial Second Level 3D Modeling Perfect Paperback.* SDC Publications. ISBN-10: 1630570389; ISBN-13: 978-1630570385.

Shingo, S. 1996. *O Sistema Toyota de Produção do ponto de vista da Engenharia de Produção.* Porto Alegre, Brazil: Bookman. ISBN 8573071699.

Simões-Marques, M., Nunes, I. L. 2012. *Usability of Interfaces, Ergonomics - A Systems Approach.* Dr.Isabel L. Nunes (Ed.), ISBN: 978-953-51-0601-2, InTech, Available from: http://www.intechopen.com/books/ergonomics-a-systems-approach/usability-of-interfaces

Sims, D. 1994. New realities in aircraft design and manufacture. *Computer Graphics and Applications, IEEE* 14(2):91, March. doi: 10.1109/38.267487

Sinha, J. K. 2014. *Vibration Analysis, Instruments, and Signal Processing.* CRC Press; 1st Edition. ISBN-10: 1482231441. ISBN-13: 978-1482231441.

Slater, P. 2010. *Smart Inventory Solutions: Improving the Management of Engineering Materials and Spare Parts.* Industrial Press. ISBN-10: 0831134011. ISBN-13: 978-0831134013.

Sobek II, D. K., Smalley, A. 2008. *Understanding A3 Thinking: A Critical Component of Toyota's PDCA Management System.* Productivity Press. ISBN-10: 1563273608. ISBN-13: 9781563273605.

Stamatis, H. 2003. *Failure Mode and Effect Analysis: FMEA from Theory to Execution.* Milwaukee: Quality Press. ISBN: 0-87389-598-3.

Stapelberg, R. F. 2009. *Handbook of Reliability, Availability, Maintainability and Safety in Engineering Design.* Springer. ISBN-10: 1848001746. ISBN-13: 978-1848001749.

Sultan, A. 2011. *Linear Programming: An Introduction with Applications.* CreateSpace Independent Publishing Platform; 2nd Edition. ISBN-10: 1463543670. ISBN-13: 978-1463543679.

Takahashi, F., Nagashima, Y., Tanaka, I., Nakada, S. 1999. Sizing and Recognizing of Cracks and Porosity in Weld Metals Using Acoustical Holographic Inspections. *Proceedings of a Technical Committee Meeting held in Kashiwazaki,* Japan, 24–26 November. International Atomic Energy Agency. Pp. 127–131.

Takahashi, Y. 1981. Total Productive Maintenance, a new task for plant managers in Japan. *Terotechnica,* 2:79–88.

Tapping, D. 2008. *The Simply Lean Pocket Guide—Making Great Organizations Better through Plan-Do-Check-Act (PDCA) Kaizen Activities.* MCS Media, Inc. ISBN-10: 0979288770. ISBN-13: 978-0979288777.

Thilakanathan, D. 2016. *3D Modeling For Beginners: Learn Everything you Need to Know About 3D Modeling!* Amazon Digital Services; Kindle Edition. ISBN: 1530799627.

Toledo, J. C., Amaral, D. C. 2001. *FMEA—Análise do Tipo e Efeito de Falha.* GEPEQ – Grupo de Estudos e Pesquisa em Qualidade. DEP – UFSCar.

Trudeau, R. J. 1994. *Introduction to Graph Theory (Dover Books on Mathematics).* Dover Publications; 2nd Edition. ISBN-10: 0486678709. ISBN-13: 978-0486678702.

Ulmer, M. W. 2017. *Approximate Dynamic Programming for Dynamic Vehicle Routing.* Operations Research/Computer Science Interfaces Series. Springer. ISSN 1387-666X. ISBN 978-3-319-55510-2 ISBN 978-3-319-55511-9 (eBook). doi: 10.1007/978-3-319-55511-9.

Villemeur, A. 1992. *Reliability, Availability, Maintainability and Safety Assessment.* Wiley. ISBN-10: 0471930482. ISBN-13: 978-0471930488.

Wang, J. 1998. *Timed Petri Nets, Theory and Application.* Springer Science+Business Media, LLC. ISBN 978-1-4613-7531-9.

Wang, L., Mohammed, A., Onori, M. 2014. Remote robotic assembly guided by 3D models linking to a real robot. *CIRP Annals—Manufacturing Technology* 63(1):1–4. Elsevier Ltd. ISSN: 0007-8506. doi: 10.1016/j.cirp.2014.03.013

Warren, J. C. 2011. *Industrial Networking.* IEEE Industry Applications Magazine, Vol. 17, Pp. 20–24.

Welch, J., Welch, S. 2005. *Winning.* HarperBusiness. ISBN-10: 0060753943. ISBN-13: 978-0060753948.

White, G. E., Ostwald, P. H. 1976. *Life Cycle Costing Management accounting,* USA, Pp. 39–42.

Wolsey, L. A. 1998. *Integer Programming.* Wiley-Interscience; 1st Edition. ISBN-10: 0471283665. ISBN-13: 978-0471283669.

Index